Farmageddon

Farmageddon

Food and the Culture of Biotechnology

Brewster Kneen

NEW SOCIETY PUBLISHERS

Cataloguing in Publication Data:
A catalog record for this publication is available from the National Library of
Canada and the Library of Congress.

Cover design by Miriam MacPhail; graphic by John Nedwidek.

Printed in Canada on acid-free, partially recycled (20 percent post-consumer)
paper using soy-based inks by Best Book Manufacturers.

New Society Publishers acknowledges the financial support of the government
of Canada through the Book Publishing Industry Development Program
(BPIDP) for our publishing activities, and the assistance of the Province of
British Columbia through the British Columbia Arts Council.

Inquiries regarding requests to reprint all or part of *Farmageddon: Food and
the Culture of Biotechnology* should be addressed to New Society Publishers at
the address below.

Paperback ISBN: 0-86571-394-4

To order directly from the publishers, please add $4.00 to the price of the first
copy, and $1.00 for each additional copy (plus GST in Canada). Send check or
money order to:

New Society Publishers,
P.O. Box 189, Gabriola Island, BC V0R 1X0, Canada.

New Society Publishers aims to publish books for fundamental social change
through nonviolent action. We focus especially on sustainable living, progressive leadership, and educational and parenting resources. Our full list of books
can be browsed on the world wide web at: http://www.newsociety.com

NEW SOCIETY PUBLISHERS
Gabriola Island, British Columbia, Canada

Table of Contents

Acknowledgments

R eaders familiar with my writing will recognize that this book is really a conversation, a conversation with so many people, some living, some dead, some friends, some strangers, that they cannot possibly all be thanked individually. But I can and do thank you all for your contribution to this work, even if you have not been aware of it.

I do, however, wish to thank a few by name who helped me directly with their critical comments on the text: Elisabeth Abergel, Jamie Kneen, Rod MacRae, Ruth Schneider; and Cathleen Kneen, for a great deal more than critical comments.

Thanks are also due to Tess Hooks, for permission to use material from her Master's thesis; to Tess, Fred Buttel, and David Kronfeld for permission to include material from their unpublished article "Scientific Conduct"; and to Ricarda Steinbrecher and the Women's Environmental Network for permission to include here, as an appendix, her excellent illustrated article on "The Science of Genetic Engineering."

I also want to express my gratitude to the men and women in universities, corporations, and government agencies such as Health Canada who have drawn a line they are not prepared to cross and are willing to take the consequences of maintaining their integrity for the sake of the public good.

May the weeds in the crops of monoculture grow strong.

Introduction

What was the problem?

We started farming in 1971. The farm we bought came equipped with a herd of beef cows and a bull. We soon sent the bull to the slaughter house and bought a sleek replacement with handsome, beefy conformation and a good family history. We were buying "genetics," though the word was never used in those days; we were still dealing in and talking about whole animals. We subsequently bought other bulls, with equally respectable family trees, as we sought to improve the quality of our brood cows as well as the quality of the calves that went to market at six months as feeders, or to the freezer trade as grass-fed beef at fifteen months.

Eventually, after similarly building up our sheep flock, we decided that we could not run both beef and sheep operations, and we sold off the cattle, including the bull of the day. We kept the Jersey "house cow" who kept us in milk and butter and provided us, through her calves, with meat for our freezer. Up to that time there had never been an issue about breeding her because there was always a bull handy whenever she was ready. Dunbar, from down the road, used to bring his house cow up for a visit now and then as well. With the bull gone, however, we had to resort to AI — artificial insemination. The cow did not seem to mind, she "took" readily, and we did not give it much thought. Dunbar stopped keeping a cow.

Now, something like fifteen years later, I think back to those events when I read about embryo transplants, genetic screening, gene therapy, and cloning. I don't like what is being done, and I ask myself, Did it all begin with that small, apparently innocuous step of getting rid of the bull and using AI to "breed" our cow? What was the problem, for which AI seemed to be the answer?[1]

This is a question we need to ask ourselves about biotechnology — the application of technology to life and the practice of treating life as technology. We have to ask this question, and many more, because society's incursions into the integrity of organisms did not stop with AI. Splitting embryos and cutting sections of DNA out of one organism and inserting them into another are other steps. Now there are people talking about — and raising money for — the "improvement" of human lives through genetic selection and modification, mirroring how our food is already being "improved" through genetic engineering. And I keep wondering, How did we get into all this? How have we come to accept what would have been unthinkable, and perhaps repugnant, only a few years ago?

On the farm, we accepted AI. It vastly increased our choice of sires. We could have our Jersey impregnated with the semen of a prize Jersey bull if we wanted a good dairy calf. Or we could choose a Hereford if we wanted a meat calf, or any other breed that might suit our whim or our business plan. Now not only do prize bulls each sire tens of thousands of calves with identical "genetics," thanks to frozen semen, but embryos sired in a lab, frozen, and shipped by air are inserted into waiting cows all over the world to produce identical "superior" calves.

The attitude towards life that lies behind these "technologies" is well expressed by a report on Embryo Transfer Research in Ontario. "Every time an Ontario cow is artificially inseminated...there is also a better than 70% chance that the sire providing the semen was himself produced by embryo transfer (ET)."[2] The report continued, "When quality is not of overriding concern, it is now fairly easy to produce low-cost embryos from eggs harvested from ovaries collected at the abattoir and fertilized *in vitro*."* More valuable donors may be called on to provide embryos once or twice a week with the ovum pickup (OPU) procedure, in which eggs are aspirated from the donor's ovaries through a special needle introduced through the vagina. That is not all. "There are circumstances under which even the eggs from run-of-the-mill dairy cat-

---

* *In vitro* is literally "in glass," i.e., in a test tube. IVF is "*in vitro* fertilization."

tle being slaughtered in Ontario could be transformed into valuable, exportable embryos by IVF with the appropriate sperm. 'Abattoir waste eggs' might be cheaply recycled into valuable embryos by tapping the ovaries and testes of fetuses early in pregnancy, when the fetuses can still be removed from the cow without affecting her fertility."

But what was the problem?

Only recently has it occurred to me that the problem was farm consolidation — called rationalization at the time — and the consequent disappearance of small diversified family farms. AI became necessary for us because there were no farming neighbors and no herd sires to which we could take our cow. It was more "rational" to reduce the bull to a straw of frozen semen stored in a flask of liquid nitrogen in the back of the AI technician's car than to load our cow on a truck and seek out a bull many miles away. Besides, frozen semen is so much more sanitary, and when it goes a step further, embryos can be screened for defects and diseases before freezing and subsequent implantation.

Birth and the establishment of life are much more complex than such "technological" operations, however. At birth the new organism leaves one context and quickly has to take on another. Even with the help of some amazing buffering mechanisms or techniques along the way, the organism is still highly vulnerable in the transition while its immune system acquires the ability to live with the immense variety of organisms already inhabiting this larger world. In mammals, including humans, of course, this process of acclimatization and immunization is carried out by means of the colostrum, or first milk, that the baby receives through suckling. There is no substitute for this process. There is no way that an "infant formula" can be made to match the precise environment into which an infant is born. Only the mother, or possibly a wet nurse already living in that environment, can provide what the newborn requires.

Business intervenes aggressively nevertheless, pushing "scientifically formulated" milk substitutes fortified with drugs — and soon, we are told, with genetically engineered (GE) additives — to replace the "unsanitary" and "inefficient" practice of breastfeeding, or of allowing calves to suckle their mothers to gain the immunities and vigour that come with colostrum.

There are parallel commercial procedures for other animals. The

US Food and Drug Administration, for example, not long ago approved a product containing a mixture of commensal intestinal bacteria isolated from adult chickens. Acting on the principle of competitive exclusion, the product is sprayed over newly hatched chicks, which ingest the waterbased bacterial soup as they groom their feathers. The commensal flora nourish and exclude pathogens (such as salmonella) in the "naive" intestinal tract of the chicks.

By the time the chicks are at least three weeks old they have a natural resistance to salmonella because the gut is already colonized by 300 to 400 different commensal bacteria. The reason younger chicks lack this protective flora, it seems, is that eggs are now hatched in sterile incubators and the chicks never know a hen until they become one themselves. Before the advent of modern poultry farms, hens hatched the eggs themselves, and their droppings, containing bacteria, were often the first meals for chicks, helping the youngsters develop immunity to foreign organisms like salmonella.[3]

The profound technological developments that have transformed the normal life of a chick, or a cow, are little different from those that have transformed the gestation of a human infant: amniocentesis, genetic screening, gene therapy, and infant formula. There is little reason for the corporate interests behind all this to distinguish a uterus with two legs from one with four. And a mammal is a mammal.

The genetic manipulation of living organisms seems to affect us less when plants are the subject than when it is done on animals or human beings. We don't seem to identify with the plight of the seed surgically divided in two as much as we might with the embryo of a cow being split to produce twins (or more), much less with the same fate being inflicted on a human embryo. The whole process is mechanized, and the actor alienated from his or her own action by the mediation of high-power microscopes.

Yet there is growing public concern about the proliferation of GE and usually transgenic plants (containing genetic material from species that would not normally cross or interbreed) such as herbicide-tolerant canola, soybeans, corn, and cotton, or the potatoes, corn, and cotton engineered to contain an alien bacterial toxin, or laurate canola containing a gene from the California bay laurel tree.

Again, I wonder where we might have gone off the track. Where did plant "improvement" cross the line into the unpredictable realm of genetic engineering, the deliberate reconstruction of living organisms to create novel life forms for commercial purposes?

On a global scale, when did the small businesses that once catered to community needs become the transnational corporations responsible for the demands of the International Monetary Fund (IMF) that national economies undergo "genetic engineering" — structural adjustment — for similar reasons of furthering corporate control and profit?

The purveyors of biotechnology and their stockbrokers are inordinately fond of saying that there is nothing new about biotechnology; farmers have, after all, been selecting and crossing plants and animals for millennia. They would claim that what we were doing with our sheep when we kept the ewes that thrived under our particular farm and management conditions, and when we chose particular rams to provide us with more breeding stock or market lambs, was biotechnology.

It was no such thing. We were not applying technology — other than the technology of pen and paper for record keeping — to the sheep in our care, and we were not imposing genetic uniformity — quite the opposite, for we found a degree of diversity necessary. Nor were we seeking — or practising — speed and precision.

Similarly, when we worked to "improve" our permanent pastures through rotational grazing and livestock management, the plants and the soil organisms were certainly undergoing changes, but they were doing so on their own terms, within their own limits. We were not violently forcing plants or microorganisms to conform to our model of what they should be in order to maximize our profits. We were encouraging transformations to be sure, but they were gradual, subtle ones that we could only observe after the fact, as with a sheep that knew how to eat, could utilize well what it found to eat, and could birth and mother a lamb with minimal interference by the shepherd.

What we were engaged in was a complex dance with the genetics, if you will, of soil organisms, plants, and animals in undefined and unending patterns of complex interaction and evolution. We were not calling the tune, though we might try to harmonize. In the course of the dance we could observe the health of the whole and at least some of its constituent elements, and nudge our partners a little this way or that — depending on how strong-willed they were. Over the years we overcame our cultural heritage of intervention and learned to observe more carefully, hoping to gain some insight into what was really taking place before our eyes and under our feet, much as farmers and herders have done for millennia.

&

> Modern science and medicine positions itself as the champion of human values, arguing that to fail to act to eliminate the vagaries of nature is to be inhumane. — Gina Maranto, *Quest For Perfection*

Now I feel it is imperative to look back (which makes me a Luddite in the views of some) and review other developments and experiences that we took for granted at the time, with the hope of gaining new insights into how we got from there to here, here being a place where the genetic manipulation and patenting of life forms are described as progress and proclaimed to be a moral imperative.

When a few years ago I wrote about the transformation of rapeseed into what is now canola,[4] I learned a lot about modern plant breeding. It is only years later that I revisit that history and find it has new meaning (like the experience with our house cow) in the light of the genetic surgery that is now being performed on canola and many other plants.

Canola is not the product of genetic engineering as the term is used now. It is actually a rapeseed — a member of the vast and highly diverse brassica family that includes turnips, broccoli, and mustard — with certain legally defined oil and meal characteristics. It was "created" through traditional selective breeding, by scientists growing out generation after generation of crosses, analyzing the properties and agronomic characteristics of each generation, and then adjusting the breeding program in the hopes of moving in a specific direction, toward particular desired traits and characteristics — essentially the same process we had engaged in with our sheep.

Keith Downey was a key player in this transformation of rapeseed, using an eye surgeon's scalpel to slice a rape seed in half. With what surely seemed like a small step at the time, though it greatly enhanced the traditional plant breeding he was engaged in, he realized that because rapeseed was a dicotyledon, each half of a carefully sliced single seed would contain the complete genetic code of the whole seed. This meant that he could set one half aside, then analyze the oil and meal characteristics of the other half. If the seed was developing in the direction of the characteristics he was after — if these characteristics were "improved" — he could then grow out the half he had set aside to produce parent stock for the next generation. This ability actually added a dimension of precision to the traditional process, but it was not genetic engineering. No foreign genetic material was being transferred, no genes were being reversed or excised, there was no direct manipulation of genetic material. The transformation was still taking place within the boundaries observed by the plant itself.

Since then, however, canola has become the darling of the genetic engineers because of the relative ease with which its genetic structure can be manipulated and because the vastness of its family provides an immense variety of characteristics that can be stolen and recombined to reconstruct canola into ever more novel plants. This, in turn, has created a dilemma for the health-conscious consumer who has been told, and has reason to believe, that canola is among the healthiest of edible oils. Now much of the canola oil on the supermarket shelf, unsegregated and unlabelled, is derived from transgenic canola engineered to resist one herbicide or another, with unknown consequences for ecological and human health.

Looking back now, I tend to think that the violent intervention of Keith Downey's scalpel — the "technology" he introduced — was, in fact, symbolically and practically the beginning of commercial genetic engineering, which has become the deliberate restructuring of life to achieve an extrinsic value or goal, or "management by objectives."

It was also an expression of a loss of respect for the integrity of the seed, a lack of respect which finds much fuller expression in current practices of genetic engineering.

This lack of respect for the amazingly complex world of life we are born into, coupled with a belief in progress achieved through ever more powerful technologies, breeds a profound dissatisfaction with life as it is. This in turn drives a relentless quest to improve on life, while *hubris* arms us with the assurance that we can do no harm, and if we do make mistakes, we will soon find another technological fix for them.

Thus if through genetic engineering of novel organisms and their release into the environment we wipe out some genetic diversity, such as a whole family of rapeseed, we are assured that genetic engineering will itself create new diversity. If, in the name of combatting some plant or human pathogen, we design and release a novel virus that turns on its creators, we are assured that it can be recaptured and brought under control with yet another novel organism.

The fact that we do not really know what the long-term consequences of genetic engineering will be, and are not prepared to move slowly and take the time to find out, means that a grand experiment is taking place, and the outcome is anyone's guess. The Bt potato, Roundup Ready canola, or Liberty-link soybean might be harmless, or they might breed a disaster. We won't know until it is too late.

On the other side of the ledger, however, is a process we do know about: organisms mutate and adapt and survive. We too can develop resistance to the restructuring of life for corporate profit.

Chapter One

Undertones of death

Don't we have a cheerful, simple morality here in Western
Civilization: expect perfection, and revile the missed mark!
— Adah, in Barbara Kingsolver, *The Poisonwood Bible*

I suspect that most of us would accept the idea that the selection and breeding of plants and animals are "life sciences," the term that major transnational corporations have now appropriated to describe their drug and genetic engineering activities. But in their hands, and labs, these activities all have undertones of death. The food, health, and environmental care that they promise, at least in their promotional materials for investors, seem always to come with the cost of violent intervention — invasions — and death for bacteria, plants, and animals, including human beings.

The suggestion that biotechnology is really about administering death may sound harsh, but consider the GE crops that have been developed by the life sciences industry. Canola, soybeans, corn, and cotton have all been genetically altered (immunized, so to speak) so that they are able to withstand lethal doses of particular agrotoxins (herbicides) aimed at anything else green that grows in their midst. The result is that the life of the designated crop is "protected" by its genetic transformation while the chemicals do their killing job on everything else. Up to now the agrotoxin industry has been surprisingly successful in dressing up and obscuring the essential lethal function of its prod-

9

ucts. The appeal of a weedless lawn, bugless garden, and unblemished fruit has been able to conceal the deaths required to achieve a uniform cosmetic appearance free of uninvited intruders.[5]

In the competitive market of the dominant culture there are winners and losers. We are encouraged to give little thought to the losers — the expressions of biodiversity that deviate from the established norm, whether they are an errant weed or a human genetic "defect." But what if it is your own child who is analyzed in the uterus as being "defective" and in need of genetic therapy or "improvement" — or one whom some life scientists judge to be a loser and not worthy of continuation, despite the gross margins for error in the analytical technology as it is. In the years — not so long ago — of Central and South American military dictatorships, the term "disappeared" was used in reference to the political undesirables eliminated by the military.

The most misleading twist is that one of the major, and most lucrative, goals proclaimed by life scientists is the prolongation of life — if not immortality itself — through elegant surgery, from genetic therapy on a human embryo *in utero* to the replacement of worn or defective — and ever more intimate — human component parts. This reconstruction of the human body, with its dependency on other animal species as the factories of these parts,* may keep the biological organism going for a few more years, like a new transmission in an old automobile; but this reconstruction is also creating a different organism, even a different person, that is no longer moving from birth to death. The refurbished organism becomes devoted to the prolongation of its biological life, perversely similar to the crops dependent on "crop protection agents" because they have been deprived of a familiar and supportive environment and transformed into aliens in a hostile land. Avoidance of death, not the fullness of life, becomes the golden rule — the gold going to the life sciences corporation.

Death, however, is an integral aspect of life. A plant dies back once it has gone to seed, that is, given its life over to the next generation. Death is 'overcome' precisely when it is taken up into life and accepted as the final act of being alive. This is a widely held religious perspective. The monoculture of industrial agriculture and, indeed, western culture and science as a whole, is built on a radically different attitude toward life and death, with its practice of administering death to 'others' —

---

*Cross-species transplants are called *xenotransplants*.

defined as 'weeds' or perhaps as 'defective' — so that an elite may survive. In contrast, the unreserved embrace of life, and its expression in organic or ecological farming, focusses on the community of the living, not on competition and killing. With long rotations in healthy soils literally seething with microbial life, mixed cropping patterns, cultivating (weeding), and careful timing of seeding, weeds lose their power and in some cases even become companions and sharers of the space, contributing to a healthy ecology, for example, by reducing water loss. Minimal intervention becomes the rule, rather than massive intervention.

Bt is a good example. *Bacillus thuringiensis* is a bacterium found in the soil in many forms. It has long been known and used as a very specific natural insecticide, applied as a topical spray when insect predation could not be coped with in any other way. Because of its specificity and biological character as a naturally occurring insecticide, it has been acceptable for use in organic farming.[6]

Now, however, a number of corporations have decided that it would be really profitable to engineer the Bt toxin into the plants themselves; then they could charge a premium for the patented seed. As a result we are faced with a number of crops — cotton, corn, and potatoes so far — with isolated, synthesized toxins from these natural organisms genetically inserted into the plants so that the toxins will be produced continuously and throughout the plants. The intent is to provide, by "natural" means, an omnipresent insecticide that will kill the corn borer, the cotton boll weevil, or the Colorado potato beetle larvae when they take their first bite into the GE plants.

Once again we have the model of the living monocultures surviving only at the expense of death to their neighbors and co-habitants (defined as pests and predators, plus, of course, the predators' predators, and their predators, and so on). This is a profoundly different model of ecology than that of competitive exclusion, or commensality and cooperation. The agricultural practices of extreme monoculture perpetuate the view of life as competition. Kill the enemies. Take no prisoners! Don't ask what created the problem or how the problem might be addressed if life itself were valued, and not just a life for the "chosen" — plants, ants, or people.

There are other approaches. For example, researchers at Cornell University think they have figured out how squash beetles ward off attacks from predators. "The squash beetles secrete their chemical defences during the week-long resting stage just prior to adulthood. When an ant encounters the chemical mix suspended in droplets on the end of tiny hairs along the pupa's yellow and black body, it scurries off and vomits."[7] This approach to the protection of life is in startling contrast to the culture of agrotoxins and biotechnology in which we seek to extend our domain by deciding who or what we want to include in it under our control, then marginalizing, killing, or destroying all the rest. The squash beetle, after all, simply wants to be left unmolested.

We must conclude that these life sciences are really death sentences. The science is devoted to ridding the world of designated enemies — designated culturally as weeds, in whatever biological form — on the basis of a philosophy that says there is insufficient room and resources for all life, that life is competitive "survival of the fittest," and that the life of some requires the death of others.

It is true that there is a ceaseless turnover of life forms, from the cells of our human bodies to the microorganisms in the soil and everything in between. But genetic engineering builds on a less selective and more deliberate elimination of life forms, a tradition expressed in the form of agricultural chemicals — agrotoxins — that sees nature as the enemy to be conquered and Creation a resource to be exploited. "Ethnic cleansing," i.e., genocide, expresses essentially the same attitude; it is simply the downside of genetic "improvement."

The small headline in the Oklahoma Farmers' Union paper caught my eye: "Death Takes Oldest Cleveland Co. Member." The farmer in question was born in 1895, but it was not his longevity that made me stop and think. It was the recognition of Death as an active player in the drama of life, a matter-of-fact statement, unadorned with sentiment. The gentleman died of old age. No heroic transplants or miraculous drugs. But he did not simply die at a ripe old age, according to the headline. Death came and took him.

In our lust for longevity, if not immortality, is there any space for "a ripe old age"? Has life simply become a technological pursuit to enhance corporate profit — a demand for food, health, and hope to be globally sourced and delivered by a Monsanto, a Cargill, a Novartis, or some other transnational corporate non-entity?

As Ivan Illich's close friend Lee Hoinacki expressed it,

> To see death today means to be able to distinguish clearly between a science-technology death and my own death. A science-technology death, of its very nature, cannot be my own death. The science-technology project, through its images and promises, practices and instruments, its aura of expertise and concern, establishes its hegemony over people's imagination and reason.[8]

&

An extreme, and logical, expression of western industrial culture was revealed in what the Rural Advancement Foundation International (RAFI) immediately and brilliantly christened the Terminator Technology in the spring of 1998. The intent of this elegant feat of genetic engineering is to produce seeds that are sterile. It is an engineered defeat of the life force — the drive to reproduce — dreamed up for the sole purpose of extending corporate control and profit.

Every spring or early summer, when we started to make hay, I was always amazed by the incredible determination of the grasses and legumes to reproduce. The trick to getting a good second cut of hay was to get the first cut before the seed heads matured. After they were cut, the plants would work feverishly to produce a second seed head before the shortening days told them the season was over. Each plant obviously knew it would not live forever, so it tried to live fully and ensure that its (genetic) line would not end in the hayloft.

The Terminator Technology wants to play a dirty trick on the plant by violently manipulating (engineering) it so that its life drive is sterilized, or dead at puberty, so to speak. Thus the "owner" of the seed gains total control of its "intellectual property" by deciding when it will die, denying the life of the plant to anyone else, including the farmer who purchased and planted the seed.

The company that received the patent for this technology has desperately tried to reclaim the high ground by referring to it as a "technology protection system" or TPS. Such terminology places the genetic engineering of sterility in the same camp as agricultural chemicals — herbicides, pesticides, fungicides, or agrotoxins as I prefer to refer to them — which were renamed "crop protection agents" a decade ago by the industry, which also changed the name of its lobby organization from the Canadian Agricultural Chemicals Association to the Crop Protection Institute in 1986. (Its US counterpart followed suit a decade later.) The intent of such linguistic changes was obviously to shift the public image from one of killing weeds to protecting food. The names and images of the agrotoxins remain, nevertheless, those of aggression and killing. For example, in Cyanamid's arsenal today are Prowl, Pursuit, Avenge, Pentagon, Sceptre, Squadron, Steel, Raptor, Cadre, and Lightning.

A spokesman for the company receiving the patent, and co-developer of the technology with the US Department of Agriculture (USDA), felt compelled to give the Terminator Technology a different meaning:

> The development of a protection system for use in self-pollinated crops is a breakthrough that will give companies a way to receive a fair return on their investment leading to future research investments and improved economic returns to farmers.

> Delta & Pine Land Company (D&PL) and the USDA are currently developing a germplasm and/or technology protection system (TPS) covered by their patent...Varieties developed incorporating this technology will allow farmers to grow a normal crop the first production season. However, seed produced and saved from this crop will not germinate the following generation and will eliminate the ability to gain multiple use from one purchase.

With stunning arrogance, the spokesman continued:

> The ability to prevent multiple use from one purchase of improved varieties of selfpollinated crops will benefit the world agricultural community by insuring that farmers in all areas of the world have an opportunity to share in the advantage of enhanced planting seed...The centuries old practice of farmer saved seed is really a gross disadvantage to third world farmers who inadvertently become locked into obsolete varieties because of their taking the "easy roads" and not planting newer, more productive varieties.[9]

Melvin J. Oliver, a USDA molecular biologist and primary inventor of the technology, was more forthright about the purpose of the Terminator. "Our mission is to protect US agriculture, and to make us competitive in the face of foreign competition. Without this, there is no way of protecting the technology."[10]

And just what was the problem? Insufficient profits due to inadequate control over life and death?

Chapter Two

Moral blackmail

Whatever they do, Westerners bring history along with them in the hulls of their caravels and their gunboats, in the cylinders of their telescopes and the pistons of their immunizing syringes. They bear this white man's burden sometimes as an exalting challenge, sometimes as a tragedy, but always as a destiny. — Bruno Latour, *We Have Never Been Modern*

Biotechnology is our greatest hope...It dramatically increases crop yields. It uses less water and pesticides, offers greater nutritional value. And, in the process, there's less stress on fragile lands and forests...Food biotechnology is already making its presence felt. It's filling consumer demand with high-quality, goodtasting food products produced in ways that are environmentally sustainable. — USDA Deputy Secretary Richard Rominger[11]

With growing intensity in recent years, the public relations advisors hired by the biotech industry have been laying a moral trip on us. They present us with a moral demand coupled with a promise — or is it a threat? — that only the biotech industry has the means for us to fulfil this demand in an ecologically acceptable way.

The demand is global — feed the world and save the environment — but the appeal is individual, in keeping with the individualistic culture of the west. There is also the subtle and pernicious assumption that it is *we* who must feed the world. There is no suggestion that this moral imperative is itself immoral, and that the people of the world might well be able to feed themselves if we would leave them alone and not demand that they produce luxury foods for us.

In May 1998, for example, Monsanto, the giant chemical company turned life sciences corporation, called upon African leaders to sign a statement telling Europeans not to be selfish by slowing acceptance of GE crops. The letter, sent by US public relations firm Global Business Access, Ltd., of Washington, DC, to carefully selected Africans, invited them as "developing country leaders" to endorse Monsanto's attached statement, "Let the Harvest Begin," which Monsanto said it intended to publish in Europe in the summer of 1998. The statement concluded, "A message from some of the world's most respected voices, made possible by some of the world's most respected companies, including Monsanto ...."

> We all share the same planet — and the same needs. In agriculture, many of our needs have an ally in biotechnology and the promising advances it offers for our future. Healthier, more abundant food. Less expensive crops. Reduced reliance on pesticides and fossil fuels. A cleaner environment. With these advances, we prosper; without them, we cannot thrive.

> As we stand on the edge of a new millennium, we dream of a tomorrow without hunger. To achieve that dream, we must welcome the science that promises hope. We know advances in biotechnology must be tested and safe, but they should not be unduly delayed. Biotechnology is one of tomorrow's tools in our hands today. Slowing its acceptance is a luxury our hungry world cannot afford.

Monsanto's offensive was met with outrage from people around the world, and by an immediate counter-campaign condemning the statement's manipulation of opinion and facts, signed by delegates of African countries participating in the United Nations' Food and Agriculture Organization (FAO) Commission on Genetic Resources in Rome at the time.

The summer of 1998 ended, and Monsanto's statement remained unpublished. Finally, in mid-October, a subtly revised version appeared in some European newspapers and on Monsanto's website, signed by figureheads and people with titles from Africa, Latin America, and some Asian and eastern European countries — fifty names in all. The changes in the text blunted the worst of the earlier tone of blackmail and the attribution of a messianic function to biotechnology. There must be at least a few Monsanto shareholders who wonder if any such propaganda campaign is really what they intended to invest in.

> Miles Russell, a public relations consultant with the notorious firm of Burson-Marsteller, told a biotechnology conference in Saskatoon that biotechnology has received an icy reception from European consumers because corporations have done a bad job of marketing. Russell, who specializes in perception management ... advised against using the argument that biotechnology is necessary to curb starvation in the developing world. That comes across to consumers as blackmail.[12]

☘

I remember being told as a child, "Clean up your plate — think of the starving children in Africa (or China or ...)." I always wondered just what the connection really was between the liver — or broccoli — on my plate and the reputedly starving Africans (or Chinese or...). I don't remember whether the guilt really worked on me or not.

In the same way, the biotech industry makes a practice of deliberately confusing personal and social responsibility, or translating political issues into personal moral imperatives that individuals cannot meet. The corporations then position themselves as the morally upright creatures that will discharge our moral responsibility for us — if we will just give them licence to do so. The biotech industry scolds that it would be irresponsible of us to stand in the way of progress and feeding the world through biotechnology — as if that were the issue. A socially constructed problem is defined as a personal responsibility.

The vice president of ICI Agricultural Products (ICI was once Imperial Chemical Industries of Britain and is now Zeneca) wrote an open letter to his industry colleagues in 1989, apparently to gather support for a campaign he was calling "A Thin Line — Feast or Famine." It started with a familiar moralism: "There is a shortage of food in the world — one billion people go to bed hungry every night, 750 million are malnourished, 15 million children have inadequate food and starve to death every year. These dramatic, but true, facts are difficult for the well fed (overfed) population of this country to perceive." (The malnourished and hungry at home are conveniently overlooked, if their presence is even perceived.)

Next, the heavy-duty moral trip: "In fact, overfed populations are very effective at: 1. Introducing legislation to reduce acreage and lower production of food which could be exported to feed starving people; 2. Reducing dependency on advanced agro-technology and encouraging practices abandoned 50 years ago (organic farming) that resulted in minimum yields and poor quality; 3. Listening to media focus on emotional (unscientific) scare tactics which could effectively paralyse agriculture" (parentheses in original).

This goes on for eleven pages, with no references and no evidence to back the outrageous claims that "Agricultural chemicals are so vital that eliminating them would mean a 40% drop in the world food supply" and "Without crop protection, food prices would jump 40-75% .... If it is immoral to use genetic engineering in some instances, it is also immoral to deliberately avoid using it in other instances .... It would be immoral for scientists to ignore the beneficial uses of genetic engineering for preservation of life and its quality on the planet."

As the following recent statement from a university professor illustrates, the party line has not changed.

> The social benefits of genetic engineering are considerable: treating human and animal diseases; increasing food production from crops and animals; increasing the nutritional value of foods; extending the shelf life of food products; reducing the need for potentially harmful chemicals such as pesticides; improving processing techniques for food and drugs; developing diagnostic tools; manufacturing cleaner fuels like ethanol to replace nonrenewable resources like oil; providing insight into the growth processes of cells (which has many potential applications, like curing diseases); and helping to provide a cleaner environment. Given these benefits, it can be argued scientists have a moral and ethical responsibility to pursue the positive uses of genetic engineering.[13]

Another expression of moral blackmail is the tactic of retreating from debate by personalizing the miracle of biotechnology. "My daughter's life was saved by [synthetic] insulin, a product of biotechnology! Would you deny her that?" You don't get far by responding that we should be considering why diabetes is so prevalent in western society.

This behavior, in reference to new reproductive technologies, is succinctly described by German scientist Maria Mies:

> Reproductive technologies have been developed not because *women* need them, but because *capital* and science need women for the continuation of their model of growth and progress...The methodological principle is to highlight the plight and unhappiness of a single individual and appeal to the solidarity of all to help that individual. In this all kinds of psychological blackmail are used.[14]

As Ivan Illich has pointed out, "By claiming [or implying] that we are responsible for the world, we also imply that we have some power over the world; and, by being convinced that we should pursue our so-called scientific endeavor of remaking the world, we enhance our need to believe that we are responsible for it."[15] Illich was not discussing genetic engineering in the late 1980s, but we can now see how literal the use of the word "remaking" has become.

Are we responsible for the world? Are we responsible for feeding the hungry? Is there validity in the threat that hundreds of millions of people will go hungry if *we* do not feed them, using genetic engineering biotechnology to do so?

And are we responsible for life and death as processes to be managed? If so, by whom and for whom are they to be managed?

Is remaking the world a good idea? If so, in whose image?

&

"Will the World Go Hungry?" asked *Time* magazine a couple of years ago. "For decades modern Malthusians have been warning of a bleak future. Sooner or later, they doggedly predict, the world's swelling multitudes will outstrip their limited food supply, and the inevitable result, as...Paul Ehrlich wrote in his 1968 book *The Population Bomb*, will be a catastrophe of horrible proportions in which 'hundreds of millions of people will starve to death'." *Time* assures us, however, that even though the population will hit 10 billion, "farmers can meet the challenge with modern biotechnology and a little bit of ancient wisdom."[16]

It is interesting, and disturbing, to see how converts are made. Plant breeder Norman Borlaug is widely recognized as the father of the Green Revolution of the 1960s and 1970s. The Green Revolution was substantially financed by the Rockefeller Foundation, with its commitment to the development and spread of capital-intensive western science and technology. The term "Green Revolution" refers to the development, by Borlaug and others, of short straw grains that are high yielding as long as they are supplied with enough water and expensive inputs such as fertilizer and pesticides.

While some have wanted to describe agricultural biotechnology as the second Green Revolution, Borlaug was quoted in the January 1997 issue of *Atlantic Monthly* as saying, "Unless there is one master gene for yield, which I'm guessing there is not, engineering for yield will be very complex. It may happen eventually, but through the coming decades we must assume that gene engineering will not be the answer to the world's food problems."[17]

However, having been a true believer in his Green Revolution, it appears that it was not too difficult for Borlaug to become a true believer in biotechnology. A few months after the *Atlantic Monthly* article was published, Borlaug made a speech in which he said, "I am now convinced that what began as a biotechnology bandwagon some 15 years ago has developed some invaluable new scientific methodologies and products..."

> Science and technology are under growing attack in the affluent nations where misinformed environmentalists claim that the consumer is being poisoned out of existence by the current highyielding systems of agricultural production...I now say that the world has the technology — either available or well advanced in the research pipeline — to feed a population of 10 billion people ....

> Extremists in the environmental movement from the rich nations seem to be doing everything they can to stop scientific progress in its tracks. Small, but vociferous and highly effective and well funded, antiscience and technology groups are slowing the application of new technology, whether it be developed from biotechnology or more conventional meth-

ods of agricultural science. I am particularly alarmed by those who seek to deny smallscale farmers of the Third World - and especially those in subSaharan Africa — access to the improved seeds, fertilizers, and crop protection chemicals that have allowed the affluent nations the luxury of plentiful and inexpensive foodstuffs.[18]

There is a disturbing similarity between the words of Borlaug, the words of Monsanto's selected African leaders, and the speeches of Monsanto executives.

&

A decade ago, population control was proclaimed as the major moral issue facing the world. In more recent years the call for population control has been redefined because it too easily sounded like a eugenics program with the goal of protecting the white northerners' affluence and privilege from the irresponsible "colored" inhabitants of other lands to the south. Better to stake out the moral high ground of feeding hungry people.

In 1995 Lester Brown, of the Worldwatch Institute, led the way in this transition with his tract *Who Will Feed China*, in which he warned of catastrophe. "An age of relative food abundance is being replaced by one of scarcity. As the one fifth of humanity who live in China seek to join the affluent one fifth already living high on the food chain, the transition into the new era will be accelerated."

Neither justice or equity enter Brown's argument. Not even charity. His final argument for feeding China is that if we don't, social upheaval will result: "food scarcity rather than military aggression" will be the threat to our security.

One of the most notable characteristics of Brown's alarmism is that in assuming the "white man's burden," he also assumes that the only course of history is for everyone in the world to become like us — white, middle-class, industrialized, capitalist, and dependent on a handful of staple crops. Nowhere in his work is there the slightest hint of recognition that for most people self-reliance or subsistence is the rule of food security, not the global market. They grow, eat, and survive on crops that are never mentioned by Brown, who reckons only on the few highly visible crops that are traded on the global market — which does not even include rice!

Recently attention has shifted away from China — since it seems to be managing rather well, for now at least — to the global population in general.* The argument remains the same: we face the moral challenge of feeding this growing population. Now, however, the solution is offered in the same sentence: this can only be done through the unrestricted application of biotechnology.

We are not supposed to notice that this is potentially a far more lucrative business opportunity than population control. Reducing the population is bad for the economy and, more particularly, bad for agribusiness, even though the same companies might profit from birth control, since they are also drug companies and sell the technology for that as well.

Once someone has been sterilized or vasectomized, market expansion is limited. Once someone has died, it is only the florists and the undertakers who benefit commercially — and then only once per death. On the other hand, there is nothing better for the Gross Domestic Product than a rising percentage of sick and dying people to be cared for who need all sorts of marketable and profitable goods and services, from drugs and wheelchairs to physiotherapists and nurses. The drug industry stands as an omnipresent reminder of the commercial benefits of a sick and dying population. Death is an essential and valuable threat. The same logic, of course, holds for plants and animals. The more dependent they can be made to be, and the longer they can be kept in that condition, the more profitable they become and the more "value" they can add to the economy.

Feeding the world, however, is a rather daunting task for the average person. If the moral demand can be transformed into the less demanding task of simply supporting the development of biotechnology, angst can be alleviated and big business can go its cynical way. Biotechnology will provide better, fresher, healthier, more nutritious, food choices. It will also provide better, healthier babies and longer shelf-life human beings thanks to embryo screening, gene therapy, and drugs individually tailored to your very own DNA.

---

* China was expected to produce 18 percent of the world's corn in 1998, second only to the US.

A good example of the role of western cultural and economic institutions in propagating the ideology of biotechnology is expressed in the 1997 report of the World Bank Panel on Transgenic Crops, *Bioengineering of Crops.*

The report suggests that the potential of bioengineered crops is unquestioned by those who are "technically competent," while risks are "perceived" by the public because it is not "technically competent" to understand that there are no real risks. Such an argument is a good indicator that the World Bank report is social program, not science. Its practitioners no more welcome or understand criticism than do the priests of any authoritarian religion: "A considered and technically competent understanding of both the potential and perceived risks of bioengineered crops," says the report, "is a requisite to their successful development and use."

It appears that Alan Irwin is speaking directly to the World Bank report in *Citizen Science.* He comments that "external criticism of science and scientific institutions is taken to imply a deficit of public understanding, rather than a need for scientific reflection and self-appraisal." Irwin continues,

> For...most of the contemporary apologists of science, *science* itself is not the problem — the problem is gaining public understanding and hence *acceptance* of science .... Within such a worldview, any problematic relationship between science and citizens must be a consequence of either public ignorance or public irrationality.[19]

The World Bank report states that "more than 1 billion people do not get enough to eat," yet goes on to state in the next paragraph that "had the world's food supply been distributed evenly in 1994, it would have provided an adequate diet of about 2350 calories a day per person for 6.4 billion people, more than the actual population."

Having identified inequitable distribution as the cause of hunger, the authors of the World Bank report go on to redefine the problem as one of production. The problem, they say, is that "agricultural production is currently unsustainable." Nevertheless, "to provide increased nutrition for a growing world population, it will be necessary to expand food production faster than the rate of population growth."

The way out of this dilemma? "At their best, bioengineering techniques are highly compatible with the goals of sustainable agriculture because they offer precision in combatting specific problems without disrupting other functional components." It is interesting, then, to note how the industry itself describes this precision.

> Monsanto's Dr. Sherri Brown, who is working on transgenic wheat, described one transformation process to the American Baking Association: "We .. .offer precision. We can change genes one at a time." She explained that particles are fired with shotgun shells at the target grain embryos. "The shotgun shell is stopped with a plate, but the DNA is not and flies with metal particles at fast speeds into the tissue culture. The DNA goes in, and if the cell has not been killed, then one in a million will have the new trait integrated into its own."[20]

In contrast to the insistence of industry propaganda that genetic engineering is precise, these shotgun techniques are widely acknowledge by those directly involved, as indicated in the following story from a daily newspaper.

> The design of Roundup Ready corn began about eight years ago ... DeKalb Genetics Corporation (owned by Monsanto Co.) used two kinds of "gene gun." The first was just exactly that, a gun, using a smallcalibre bullet coated with DNA and fired into a clump of corn cells grown in the lab. It was literally hit or miss technology, but it worked: In the blast, some of the genetic material got inside individual cells. Later the technique became more sophisticated. The "bullet" is now a tiny fleck of gold, just one micron (one millionth of a metre) in diameter, and blown by a blast of air at the mass of cells. As with the .22-calibre bullet, the gold flecks have a coating of DNA. According to Ken Kasha, professor of crop science at the University of Guelph, "It will enter the cells ... and is taken inside the cells. Once it's in the cells we don't know exactly how it works. Most likely it gets into the nucleus of the cell where the corn chromosomes are" ... "It's not a very exact science," says Mike McGuire of DeKalb Canada.[21]

New Scientist magazine commented pointedly on the randomness of genetic engineering:

> At the moment, gene therapy relies on viruses and other "vec-

tors" to shuttle DNA carrying healthy genes into patients' chromosomes. But they insert the genes at random points. So the new DNA can disrupt the way normal genes work, and the new genes may not churn out proteins properly because they are not surrounded by sequences that normally turn them on and off.[22]

Clearly the assertion that genetic engineering is neat and tidy is ideological, not scientific. This is without even considering the potential social disruption that the transformation of agriculture from the outside brings with it. Unpredictable disruption of everything from DNA to ecosystems is one of the major issues in genetic engineering. As biologist Mae-Wan Ho puts it, "To understand why genetic engineering biotechnology is so inherently hazardous, we have to appreciate the prodigious power of microbes to proliferate, the protean promiscuity of the genes they carry, and their ability to jump, spread, mutate and recombine."[23]

&

If the moral imperative to feed the world is of dubious merit, so perhaps are the figures for population growth upon which the imperative is based. No references or documentation other than carefully chosen extrapolations are ever provided to back the threat of a world population of 10 to 12 billion. This is to be expected in a right-wing business journal such as *Forbes,* so the following language in an article extolling Monsanto and biotechnology is not surprising. "As mankind extends control over the environment, fears of famine fade, even though world population grows by 800 million per decade and is expected to double, to 11 billion, by 2100."[24]

In the face of such dire predictions, more solid figures can be a bit startling.

The United Nations reports that more than fifty countries, including China, now have fertility rates below replacement levels, which is conventionally put at 2.1 babies per female (the .1 allows for childhood deaths and the slightly larger number of boys born than girls). The UN's prediction of the likely world population in the year 2000 has declined from an estimate of 6.26 billion made in 1992 to a 1996 estimate of 6.09.[25] "With just 18 months to go, we are probably still well under 5.9

billion," says *New Scientist* in its review of the UN report.[26] According to the UN Population Division, the world population will peak in about 2040 at 7.7 billion and then go into long-term decline, dropping to 3.6 billion, less than two-thirds of today's global population, by 2150.

World population was about 150 million 2000 years ago, and 350 million 1000 years ago. The fourteenth century Black Plague reduced the population by a third, and by the nineteenth century and the industrial revolution, expansion was well underway. The average fertility rate worldwide peaked in 1950 at 5.0, but it has dropped since then to 2.9. In Europe it is now 1.4 according to Fred Pearce in a special "Inside Science" insert in *New Scientist*.[27]

Whatever the figures actually turn out to be won't really matter. The biotech industry has no intention of feeding anyone who cannot pay well. But the hungry and deprived can be used to prey on the guilt of the affluent so the corporations can get their way with the politicians and the regulatory agencies, get new products to market, keep the industrial farmers of the north on the technology treadmill, and make their investors happy.

Chapter Three

A case of bad attitude

"So, are you against *all* biotechnology?"

I am frequently asked this question after I have given a talk on biotechnology. It is a question of puzzlement, the questioner not being sure if they have understood me. It is sometimes a question of disbelief that anyone could be just plain against "progress." Often the questioner is engaged in a quiet personal struggle against hopelessness and despair in the face of the dominant culture of determinism and individualism that drives the practice of biotechnology.

In effect, my answer is Yes, I am against all biotechnology. Not on principle, but because, as an artefact of society, an expression of a particular culture, I think "modern biotechnology" is a bad attitude — a bad attitude towards life, towards Creation, towards other cultures and other ways of knowing and experiencing the world.

Modern biotechnology — genetic engineering — is an assault on life; not an altruistic exercise in curiosity but a demand to control.

Genetic engineering is an expression of ingratitude and disrespect, if not contempt. It is a vehicle, in practice, of an attitude of domination and ownership, as expressed in the assumption that it is possible, reasonable, and morally acceptable to claim ownership over life. The claim that it is possible to own life, at least to the extent of being able to claim a patent on a life process or life form, is so outrageous socially and ethically as to be hardly worth debating.

This attitude presumes that life is an object, a thing, a commodity, a product, and hence definable and patentable. Making a fetish of life is the way Ivan Illich has referred to this attitude. He describes it as a western notion, "a new kind of entity ... a new kind of social construct ... spoken about as something precious, endangered, scarce. It is further spoken about as something amenable to institutional management .... 'A life' is amenable to management, to improvement."[28]

This is neither an historic nor universal conception of biological life. Philosopher Michel Foucault pointed out that, "Historians want to write histories of biology in the eighteenth century; but they do not realize that biology did not exist then ... and that if biology was unknown, there was a very simple reason for it: that life itself did not exist. All that existed was living beings."[29]

In fact, as an identifiable experimental science, biology was just emerging a century ago, and the term, and practice of, "molecular biology" only came into being in the mid-1930s.

ॐ

Novartis, the company created by the merger of the Basel twins, Swiss drug companies Ciba Geigy and Sandoz, put out a series of ads in the spring of 1998 describing the company as "the world's largest life sciences company." A line common to all four ads was "new skills in the science of life." One ad was for new medicines, another for "new ways to protect crops," a third was for "new therapies to make organ transplants more successful." A slightly different ad asked the question, "How does a global corporation enhance life?" You can guess my response!

The spate of ads for life sciences corporations continued with the French drug company Rhône Poulenc, which put a full-page ad in a Toronto paper in June 1998, "Rhône Poulenc — uniting sciences for life — to improve life, we explore all of its forms."

As Illich commented a decade ago, "The ominous power of modern institutions consists in their ability to create and to name the social reality which the institutions' experts need as the substance they manage."[30]

The most ambitious claims are made by Monsanto. In the spring of

1998 Monsanto relaunched itself as a life sciences company that would "address the health needs of a rapidly expanding world" and that was "dedicated to helping people everywhere live longer, healthier lives." Monsanto's assumption of the mantle of global savior and its subsequent behavior illustrate both "the ominous power of modern institutions" and the compulsion of these institutions to "name the social reality."

The social construct of biotechnology that Monsanto has both seized and built upon is, as molecular biologist Steven Rose puts it, the attitude that there are genes available to account for every aspect of our lives, including genes that explain "the social inequalities that divide our lives along lines of class, gender, race, ethnicity." This provides the ideological foundation on the basis of which genetic and phamacological engineering can "hold out hopes for salvation that social engineering and politics have abandoned."[31]

It would be hard to come up with a more extreme — or blasphemous — expression of the compulsion to define and control the social agenda than the logo unveiled by Monsanto after splitting off its traditional chemical interests into a separate company named Solutia. With its new logo — "Food • Health • Hope" — the company obviously seeks to transcend all other possible pretenders to power with the presentation of itself as the Holy Trinity, savior of the world.*

This benign entity will feed the hungry, heal the sick, and bring hope to the distressed. This is not a role that its detractors have cynically assigned to the company, but a role that it has assumed for itself — perhaps on the suggestion of one of its hired interpreters and advisors such as the global PR firm Burson-Marsteller.**

This hubris, this overbearing pride, would suggest that nature is inadequate and hostile, an alien to be subdued, exploited, and finally

---

* *Western Producer* reporter Barry Wilson described the president of Monsanto Canada as "a biotechnology apostle" with a message of "a mixture of science as the savior of the planet and science as lucrative corporate opportunity." (8/10/98)

** In 1997 the pan-European biotech industry consortium EuropaBio had Burson-Marsteller draw up a battle plan for the biotech industry in the face of growing public hostility. It recommended "basic strategy disciplines," including the use of "Symbols - not logic: symbols are central to politics because they connect to emotions, not logic. Adversaries of biotechnology are highly skilled in the cultivation of symbols eliciting instant emotions of fear, rage and resentment. Bioindustries need to respond ... with symbols eliciting hope, satisfaction, caring and self-esteem." (This document was obtained by Greenpeace Europe and broadcast via e-mail.)

discarded in a sanitary landfill. So we proceed to intervene, rearrange, throw away the "useless" bits and the "junk," including "junk DNA," as if life itself — the organisms of every size, shape, and duration — were worthless in and of themselves until we "add value" to them. The interventions are not regarded as violent because the objects are not life. A child playfully constructing objects out of Lego is not engaged in violent activity. But what if it is a beetle or the family dog that is taken apart, or baby sister? In a culture of domination and control, how is the child to know the limits, the boundaries of acceptable intervention and reconstruction?

"Miracle products are coming — products for a good life, products which can control and perhaps even eradicate major threats to health and well being," Richard Mahoney, then ceo and chairman of Monsanto Company, told the Executive Club of Chicago back in 1993, about the time Monsanto was setting out on its divine mission. But, he warned, "Biotechnology sometimes gets put in a special class of risk by the public — open to exploitation by the growth and influence of anti-science organizations who can scare the public in the process...We have not done an outstanding job of communicating with the public — but we are learning — and we're learning fast."

Five years later, Monsanto Europe's information manager, Jonathan Ramsey, wrote, "Monsanto believes that the issues involved are far too important for the public to be left at the mercy of such misinformation and 'voodoo theories'. Whilst in the past we answered specific criticisms, we have now taken the much wider policy decision to increase the information flow to the general public and to engage in an ongoing constructive dialogue. Our aim is no longer simply to counter the obscurantism of our critics but, positively, to give factual and detailed information to consumers on subjects they are concerned about, in a language they can understand."

During the summer there was an outbreak of public rage against GE crops and biotechnology companies all over Europe and the UK. Transgenic crops were danced on, cut down, pulled up, and squatted on by locals and roving bands of protesters who recognized that neither the governments nor the corporations were the least bit interested in public opinion.

Referring to the crescendo of these "decontamination" actions in the UK, Ramsey, in typical Monsanto fashion, tried to stake out the moral high ground, not by dealing with substance but by name-calling. "There has ... been sporadic vandalism, targetting biotech field trials...This is a very different campaign indeed. It is led by a small group of so-called and self-proclaimed 'ecowarriors' - rather a grand title for a group whose primary activity is the wanton destruction of private property."[32]

Ramsey's identification of the "destruction of private property" as the greatest moral sin displays Monsanto's cosmology: private property is sacrosanct and its protection (as in "crop protection") is paramount, not life, not Creation. The subtleties of ethical behavior and the principled action of civil disobedience — "civil obedience" as Ramsey's "vandals" describe their actions — certainly find no place in Monsanto's collective conscience.

"Monsanto's challenge is to improve the world while increasing corporate profits," reported the magazine *Business Ethics* after interviewing Monsanto's new chief executive officer Robert Shapiro in 1996.[33]

Fortunately , not everyone agrees that western science in general, and biotechnology in particular, have some unique and universal authority as "the truth," or as the only correct way to view and know the world, as teacher Godfrey B. Tangwa of Cameroon points out.

> In contrast to the Western anthropocentric-individualistic outlook, the precolonial traditional African metaphysical outlook is ecological/biological-communitarian. Within the African worldview, the distinction between plants, animals and inanimate material, between the sacred and the profane, matter and spirit, the communal and individual is a slim and plastically flexible one. Similarly, metaphysical conceptions, ethics, customs, laws and taboos form a single continuum ....

> The western world has the penchant for presenting its vision, ideas, convictions and practices as universal imperatives of rationality or morality which ought to be binding on all.[34]

&

So, am I against *all* biotech?

Maybe there are bits and pieces of biotechnology that I could find acceptable, socially and ethically, but I cannot take just the pieces I like and ignore the rest, like picking the raisins out of an oatmeal cookie. In this sense I cannot be a fundamentalist, using those bits of whatever scripture endorse my predisposition while ignoring the other words that are difficult or that challenge my comfort (the "hard sayings of Jesus," for example, as they used to be described in New Testament scholarship — sayings such as "let the dead bury their dead.") No, I have to deal with the whole opus and look at its extreme expression to observe its real character — its "bottom line," to use a more vulgar contemporary term for the net results.

The underlying cultural assumptions of biotechnology and genetic engineering are lucidly articulated in the brief paragraph from a Swedish "environment business" magazine that identifies life as "word," "code," "information," and "software" and assigns suppression and control as the purpose of managing this "software."

> As coded information, genes provide a link with the philosophy of information management. Corporate gurus see a shift in the way that humanity's ongoing project of control over nature will work: instead of brute physical suppression of natural processes, a kinder process of control, through rewriting of "nature's software," is envisaged.[35]

What I see expressed in the project of modern biotechnology as a whole, as well as in its components and processes, is not a devotion to life, but a preoccupation with death and dying that allows death to define life, or the reduction of life not to a complete spoken or breathed word, but to the letters of an alphabet — information, supposedly, but devoid of life, as described in this comment on a research project.

> Freeze-dried mouse spermatozoa are all motionless and dead in the conventional sense. When injected into oocytes, however, their nuclei can support normal embryonic development even after three month preservation in a dried state.[36]

> The sperm could be stored at 4° C or at ambient temperature before reconstitution with the simple addition of water.[37]

As the freeze-dried and quite dead mouse sperm proclaim, DNA is not life, however essential it might be to life.

The underlying argument of the biotech industry is that Mother

Nature is incompetent — in the same way that women are incompetent to nurse their babies. Infants, and by inference their mothers, will be better served by becoming dependent on transnational infant formula manufactuers than by feeding themselves at mother's breast.

&

I might like or want to believe in the talk of saving lives, curing diseases, and feeding the hungry. But in examining the subject and putting it in context, its attitudes of conquest and control by means of threatening and administering death become all too obvious. As I argued in Chapter Two, it is death to microorganisms, death to larger pests of both plant and animal varieties, death to human diseases, human communities, even whole cultures.

"Gang up on the intruder" and "Kill the offending organism" says University of Guelph's head of research communications, Owen Roberts, about fighting diseases with vaccines.[38]

It is not death by massacre. It is the more subtle and pernicious death by genetic obliteration, cultural conquest, and enslavement. It is death through deprivation of culture, knowledge, and self-reliance. It is domination through the construction of dependency. The capture, transport, and enslavement of Africans to provide expendable labour to the industrializing United States comes to mind. Overt slavery is no longer acceptable, but a more insidious mechanism is being developed to enslave the entire population of the globe: giant transnational corporations are using patenting and genetic engineering to achieve control.

&

As a boy I loved to play in the streams and woods behind our house, or on the shore of Lake Erie — it was clean enough at the time, though we sometimes had to remove the dead fish — watching the water, watching the bugs, observing. I intervened with little dams or bridges made from twigs and stones, and I dug new channels, particularly during the spring thaws. I caught butterflies at summer camp and mounted them — it was a challenge to keep them whole and undamaged — so that I could admire them. I was not, at the time, bothered by catching them with a net and putting them in a jar with a bit of cotton soaked in (probably carcinogenic) carbon tetrachloride to suffocate them, though I had no desire to dissect them or any other creature (including the dead skunk I once brought home and hung out my bedroom window).

The line I drew is recognized historically by science editor and teacher Jon Turney in his account of public attitudes in the middle of the nineteenth century when vivisection (the dissection of live animals) became "an integral part of an interventionist, empiricist approach to biological problems" practised by "a new breed of physiologists [who] asserted that life could be explained in purely physico-chemical terms."[39]

I haven't changed all that much. I don't like picking wildflowers — they won't last, not having been bred or engineered for a long shelf-life, and they look much prettier in their own habitat (context) than in my constructed one.

When we started farming, with no academic or practical experience, observation was our primary learning activity. The interventions and projects undertaken without long observation were usually not too successful — if not downright disastrous. For example, I started with the deep-seated feeling that I was not really farming until I plowed a field. I don't know where that came from, and I don't know what I thought farming really was, if not all the other tasks and pleasures that filled day after day. But it was a profound inherited cultural attitude.

Unknown to me, up 'til then I had been shaped by the culture of America to define the essence of farming as ripping open the land — raping Mother Earth, to put it more bluntly. (Untilled land has long been referred to as "virgin," as in "virgin prairie," referring to the vast open expanses of the west before settlement and ownership by Europeans. Homestead claims were often contingent on "breaking" the land.)[40]

Over the years we learned to observe, to learn from our working sheepdogs and the Jack Russell terrier, from the sheep, from the exuberant water at various times of the year, and from the grass, trees, and seasons themselves. We learned to be very cautious about intervening. Perhaps the hardest lesson in that regard was in connection with lambing. Heroic measures to save a puny lamb usually turned out to be a big and expensive mistake. We should have taken our cue from the ewe, the lamb's mother, if she chose to ignore the newborn. It almost always turned out that she knew something instinctively that it took us weeks or even months to learn, after we had spent countless hours trying to salvage a weak lamb only to have it remain an unthrifty runt that could never make it to market.

So I have grown to prefer the science of observation to the science of intervention. The science of intervention seems so much more interested in achieving control than in achieving understanding.

Evelyn Fox Keller tells a wonderful tale about the life and work of Barbara McClintock, a pioneer biologist who observed in preference to intervening. Writing in 1982 about McClintock's work in the 1960s (which had begun well before she received her Ph.D. from Cornell in 1927), Keller says, "As she had long since discovered, if one looked hard enough and carefully enough, a single organism would reveal its secrets. It would tell you not of one but of many mechanisms it had evolved to regulate the expression of genes — mechanisms that enabled its cells to produce exactly what was needed, when it was needed. It was an exquisitely balanced timepiece that seemed to be capable of whatever readjustments circumstances required. Some mechanisms involved massive reorganization of the genome; others merely modulated the expression of genes without changing the DNA composition. Though she couldn't provide molecular explanations for any of these events, she could plainly see their effects. There was no question they occurred."[41]

Current advocates of genetic engineering like to praise the speed and "precision" of their technology. Keller comments that in the early 1930s McClintock made use of the newly discovered mutagenic effects of x-rays on genetic material and was pleased with the speed with which effects could be studied, a difference comparable to that between working with *Drosophila* (fruit flies) and working with corn. "If *Drosophila* had the advantage over maize of yielding a new generation every fourteen days, instead of annually, then bacteria were incomparably better. A bacterium divides in two every twenty minutes."[42]

Later on, however, McClintock decided that even growing two crops of corn in one year provided her with more material than she could properly observe and analyze. Not only that, she also found that "as she grew older, it became less and less possible to delegate any part of her work; she was developing skills that she could hardly identify herself, much less impart to others."[43] In fact, as Keller points out, "she had always carried out the most laborious parts of her investigations herself, leaving none of the labor, however onerous or routine, to others .... For McClintock, more than pride was involved. Her virtuosity resided in her capacity to observe, and to process and interpret what she had observed."[44]

Now this work of observation is increasingly delegated to robots and computers, as in the case of gene sequencing, raising interesting questions about what is being observed. Perhaps it is the consequent blindness that enables genetic engineers to make groundless assurances as to outcomes and safety.

Not long ago I was fascinated by a discussion of transposons, or transposable elements,[*] in an article about the genetic engineering of insects.[45] The article pointed out that the genomes of animals are literally crammed with transposon sequences, and "over 10% of the human genome is made up of transposon DNA, mostly fossilized relics of active transposons." Are we really a kind of walking museum I wondered? But when I read about engineering mosquitos into pest managers instead of disease carriers, or trying to improve the quality of silk produced by a silkworm, I was struck less by the science than by the underlying attitude expressed in the article: a moral certainty that such intervention is perfectly reasonable and that all Creation should be subject to modification by us to meet our demands, desires, and specifications. This is really the issue with recombinant bovine Growth Hormone (see Chapter Six): is it reasonable, ethical, or even fair to inject a cow with a drug that forces her to produce more milk, thus attacking her health and depriving her of any essential autonomy or integrity?

It's an attitude, again, that shows no respect, no gratitude. It is disturbing. Are there no limits? Is there no satisfaction?

As an artefact, a construction of the culture of the west, it seems that genetic engineering biotechnology is trying to prove that there are no limits, no boundaries. All life is fair game for control and exploitation — or eradication.

---

[*] Transposon: "A DNA sequence which can become incorporated into any DNA molecule in the cell. Transposons may be transcribed and translated. They may inactivate or modify genes with which they become associated." *(Facts On File Dictionary of Biology)*

The folly of eradication, however, is that the good guys may not always win.

To assume when a peasant tells you that she is conversing with the soil or the wind that she is speaking metaphorically is to assume that Andean peasants are the intellectual heirs to the Reformation and the Scientific Revolution, in which nature does not speak directly, but can be interrogated in the laboratory through experiments .... Just as we humans speak, so do the other inhabitants of the world. To hear these other inhabitants speak, no special training is required, just attention and practice.[46]

Chapter Four

Where Did biotechnology come from?

The term "biotechnology" itself has recently been held by the Federal Court of Canada to be a term of unsettled meaning (Astra AB v. Aastra Corp.).— *AgBiotech Bulletin*, December 1996

The most helpful definition of biotechnology I have yet found comes from Klaus M. Leisinger of the Ciba-Geigy Foundation in a footnote to a 1995 paper.[47] Biotechnology is "the integrated application of biochemistry, microbiology and process technology with the objective of turning to technical use the potential of microorganisms and cell and tissue cultures as well as parts thereof."[48]

Biotechnology therefore, according to Leisinger, "deals with the utilization of biological processes in technical operations and industrial production. Gene technology is a means to an end, inasmuch as it allows the properties of microorganisms to be modified in such a way that a desired effect is brought about through biological processes."

The Canadian government ministry Agriculture and Agri-Food Canada (AAFC) has been content to use a substantially more simplistic definition of biotechnology: "The applied use of living organisms, or their parts, to produce new products," while the Canadian Environmental Protection Act uses an only slightly more elaborate version: "The application of science and engineering in the direct or indirect use of living organisms or parts or products of living organisms, in their natural or modified forms."

At times AAFC gets a bit carried away with itself (or by the biotech industry), as in this definition of a living modified organism (LMO): "A plant, animal or microbe whose genetic material has been changed to make the organism more useful to humans. — Agriculture in harmony with nature."[49]

A more careful and complete definition of both biotechnology and LMO was proposed by the Biosafety Working Group of the International Biosafety Protocol at its August 1998 meeting in Montreal. "LMO means any living organism, obtained through the use of modern biotechnology, containing a novel combination of genetic material. Living organism means any biological entity capable of replicating or transferring genetic material, including sterile organisms, viruses and viroids. Modern biotechnology is a set of in vitro [nucleic acid] techniques that overcome natural physiological reproductive or recombination barriers, other than traditional breeding and selection."

Then there is a definition offered by Monsanto in its 1996 annual report to shareholders: "A short list of words for a life sciences company — biotechnology: Biotechnology harnesses the metabolic potential in living systems. Plant biotechnology, an extension of traditional plant breeding, is the addition of selected gene traits to plants to develop new varieties;" or the profoundly simplistic definition found in Monsanto's 1998 "Let the Harvest Begin" propaganda: "Biotechnology is the science of changing the genetic makeup of seeds that grow our food to add new benefits."[50]

Molecular biologist and organic gardener Martha Crouch defines genetic engineering "as the process of manipulating the pattern of proteins in an organism by altering genes. Either new genes are added, or existing genes are changed so that they are made at different times or in different amounts."[51]

None of these definitions, of course, describe traditional practices in animal or seed selection and breeding as discussed earlier.

What biotechnology is clearly depends on who is defining it and for what purposes. In other words, it is a matter of context and perspective.

Let us go back to the useful observations of Klaus Leisinger, who distinguishes three generations of biotechnology. "In the first, bacteria or yeast, for example, were used in making cheese or beer. In the second, microorganisms were used to produce antibiotics, and molecular biology was further developed. In the third generation, finally, it has become possible to alter the genetic material of an individual cell directly."

Are these three generations just a logical and harmless progression from traditional brewing and cheese making, or are there radical discontinuities between them? First, we have to realize that Leisinger's three generations have an identifiable context. That is, Leisinger himself speaks from a particular context as a male, a European, an employee of a transnational corporation, and a person working with a particular understanding of science. There is, however, no hint in Leisinger's comments that he is aware of and acknowledges this particular — and peculiar — context out of which he speaks and the epistemology (see Chapter Ten) that characterizes it.

Now add one or two characteristics of this culture, such as its belief in progress and in the neutrality of technology, and Leisinger's description is almost inevitable. But this culture is not universal, however much it presumes to be. It is quite possible to step aside, or back, to gain a different perspective on the subject. In doing so, we can see that what we now know as biotechnology could have taken a different direction. Power and money made the decision, not democracy or science, as feminist critic of science Sandra Harding points out.

> During the last century, the social use of science has shifted: formerly an occasional assistant, it has become the direct generator of economic, political, and social accumulation and control. Now we can see that the hope to "dominate nature" for the betterment of the species has become the effort to gain unequal access to nature's resources for purposes of social domination.[52]

Robert Bud of the Science Museum in London expresses a similar perspective in his comprehensive history of biotechnology, *The Uses of Life*. He begins his tale early in the nineteenth century, describing in detail how "physiological processes, be they fertilization or digestion," were radically reduced to the transformation of chemicals." He cites the synthesis of urea by Friedrich Wöhler in 1828 as a "useful benchmark in the erosion of the distinction between natural and chemical products."

Almost a century later a bacterium was discovered that produced both butanol and acetone from starch in a fermentation process. An inexpensive source of butanol was the key to producing synthetic rubber. "Profiting from both the technology and the public's willingness to believe," says Bud, the scientists involved in this work in Britain, "established a fascinating precedent for the future of biotechnology" by forming a company to develop the process for Britain.

The project was discussed in an article by Henry Armstrong, "The Production of Rubber: With or Against Nature?" in 1912. "We are competing with Nature in many directions at present and it is very desirable to discuss whether in the future it will be either desirable or possible to work so much against her .... Ethically we shall probably be making a mistake in not availing ourselves to the full of the activity of the plant; but, apart from this, it may be that, when everything is taken into account, the plant is able far more effectively than man to make rubber from starch."[53] At this point, Bud suggests, fermentation technology (zymotechnology; zyme = leaven) and agronomics were coming to be seen as exclusive alternatives; either the plant was seen as a factory or it remained in its context as a whole organism.

Later Bud suggests that the modern use of the term "biotechnolgy" was launched in 1961 when the *Journal of Microbiological and Biochemical Engineering and Technology* changed its name to *Biotechnology and Bioengineering*. The magazine's founder and editor, Elmer Gaden, had envisioned two major areas of microbiology: first the "long-established, traditional technologies" of "extraction, separation, purification and processing of biological materials;" and second, the "utilization of complete biological systems (e.g. cells and tissues) or their components (e.g. enzymes) to effect *directed and controlled* chemical or physical changes" (emphasis in original).[54]

In 1970, writes Bud, "the most appropriate use for continuous fermentation...seemed to be the cultivation of hundreds of thousands of tonnes of single-cell protein to feed the starving .... The post-World War II generation of idealists diagnosed new possibilities in the complex of techniques of fermentation, enzyme technology, and the processing of microorganisms that, increasingly, they called 'biotechnology'. It seemed particularly well suited to otherwise disadvantaged developing countries, rich in biological raw materials and in great need of products such as fermented foods, fuel alcohol and biogas for energy, and nitrogen-fixing bacteria. The technology would be deployed by small enterprises meeting local needs, not by remote multinationals."[55]

Business and nationalist interests, however, shifted the focus of biotechnology development from the bodily needs of the people of the Third World to the animals of well-fed Europe. "Visions of bounty that did survive encountered a new cynical strand of thinking that could be caricatured," says Bud, "as 'the use of poor countries' genetic resources to solve rich country problems' .... By the 1980s, biotechnology came to be widely seen as another high-technology threat to the Third World."[56]

This shift of interest from using biotechnology to address the problems of the poor to developing products for the affluent is particularly well illustrated by what happened to single-cell protein. Although British Petroleum was ready by 1962 to build a plant in southern France to produce single-cell protein as human food and even advertised it as such, the product was, instead, launched as animal feed for the developed world.

As recently as the late 1970s, says Bud, biotechnology as genetic manipulation was primarily concerned with genetic enhancement rather than genetic engineering. The departure from the potential development of zymotechnology was clinched, however, with the discovery of the double helix (of DNA) and the development of cut-and-paste recombinant techniques. Genetic engineering came to be defined as the basis of biotechnology. The transition was from "the concept of molecular biology as a science to the technological concept of biological engineering," or "technologically effective biology," from dealing with whole organisms to reconstructing what were seen as the instructions for the parts of life, the genetic codes of DNA.

This transition was described quite differently by a biotech business executive in 1985: "Biotechnologists had been talking of a range of hypothetical benefits ... while their critics were worried about very real risks. By 1984 ... biotechnologists were talking about 'very real' benefits, while any discussion of risks is hypothetical at best."[57]

With a note of sadness, Bud comments, "The logical leap from single tool to economy-reviving industry was improbable: nonetheless, the role of other skills was relegated to obscurity."[58]

Now we can read in government pronouncements, such as the government of Canada's announcement of a "New Biotechnology Strategy," that biotechnology is to be the engine of an economic revival.[59]

Theoretical physics and molecular biology are also strands of the political history of biotechnology, as noted by Evelyn Fox Keller, professor of history and philosophy of science at the Massachusetts Institute of Technology. As the new field of molecular biology developed, she writes, its leading figures came not from biology or classical genetics, but from biochemistry and physics. She cites Max Delbrück, a theoretical physicist who became one of the founders of molecular biology, who was "steeped in a tradition that seeks understanding in simplicity rather than complexity — that proceeds by isolating and investigating phenomena in the simplest form and treats the variety and plenitude of nature as a distraction, something to be cut through or cleared away in pursuit of general laws. As a physicist, he sought the simplest possible organism available for analysis. And in biology, simplest is likely to mean the smallest. That this might mean bypassing some of the very complexities geneticists had been trying to explain was not a problem; it was a virtue."[60]

The name "molecular biology" itself was invented in the mid-1930s by Warren Weaver of the Rockefeller Foundation "as part of a coherent policy by one of the major fundgivers in the field...As one of the early directors of the Foundation expressed it bluntly in 1934, its policies 'are directed to the general problem of human behavior, with the aim of control through understanding'."[61]

The Rockefeller Foundation, says Jon Turney, "pledged a large sum to an expanded programme of fundamental biology...to be concerned with rigorous, physico-chemical, quantitative research, related only obliquely to medicine...Here began...the path to an ability to manipulate life far beyond anything previously foreseen."[62]

"The sheer power and scale of the Rockefeller vision," backed by the vast Rockefeller wealth, continues Rose, "ensured that alternative understandings of biology withered. That was the fate, for example, of the 1930s Theoretical Biology Club in Cambridge, England, centred around Joseph Needham, whose nonreductionist approaches to metabolism, development and evolution were swept aside by the

Rockefeller offer to fund an explicitly reductive biochemical research program. Of course, the Rockefeller vision has been immensely productive in both scientific knowledge and technologies...but to naturalize it as if it were the only way of understanding the living world, and to ignore its explicit goals of social control and its implicit eugenic agenda, is to fail to grasp the directions in which it is leading us."[63]

The deliberate construction of the foundations of modern biotechnology to serve certain interests and not others has also been noted by molecular biologist Philip Regal. The Rockefeller Foundation, he says, promoted the idea that "society should wait for scientific inventions to solve its problems, and that tampering with the economic and political systems would not be necessary." The effect of this, he points out, was to nurture resignation to genetic fatalism while at the same time raising "calming expectations: science and technology would bring grace."[64]

It was, of course, this same Rockefeller Foundation that defined and financed the Green Revolution, with its overt moral agenda of increasing food production masking its social and biological agenda of extending the breadth and depth of domination and control by advanced capitalist countries and their corporations.

Nelson Rockefeller and the Rockefeller Foundation were also the primary architects of the global strategy to solve the larger problem of "underdevelopment" ("inadequate life" should we say?) by increasing food production so that the targeted countries could become food exporters and "trading partners for the free world and provide opportunities for investment."[65] Wherever implemented, the goal was to integrate the region with the mainstream US economy, not necessarily to enable people to grow more food for themselves and their families.

This was combined, in the Green Revolution, with the attitude that, "if there is not enough existing knowledge available to deal with a problem [such as a shortage of food], such knowledge could be paid for by the [Rockefeller] foundation and produced by scientists [and] embodied in the technologies [seeds]."[66]

Once the problem of hunger had been defined as one of inadequate yield per hectare, not inequitable distribution, it was to be solved by means of "improvement" in the seed. The result might be self-sufficiency or even surplus production for export in the national accounts, but it was also one of escalating entrapment for both the farmer and the state. The Rockefeller Foundation philosophy of development sought self-sufficiency in terms of output, perhaps, but only when combined with dependency on expensive inputs such as irrigation, seeds, fertilizers, chemicals, and mechanization that in most cases had to be purchased not only externally, but specifically from corporations in those countries making the development loans and grants for the modernization in the first place.

In light of this, the new-found altruism of the drug/biotech companies in offering "new skills in the science of life" (the tag-line in the Novartis ads) is transparent as a continuation of the same program of creating dependency on purchased "inputs," now not just fertilizers, agrotoxins, and improved seeds, but genetically engineered components and "precise" genetic modifications as well.

⅋

The speed of change, the availability of resources to fuel it, and the evangelical promises of some of its advocates, combined with a growing realization of the power of this new "technology," obviously disturbed some of the more thoughtful scientists and researchers and produced a mood of caution in the budding field of molecular biology in the early 1970s. A committee chaired by Paul Berg of Stanford University called for a moratorium on certain lines of research. The committee felt that not enough was known about the possible outcome of some genetic manipulation experiments, particularly "those which might confer novel antibiotic resistance or the ability to make toxins on strains of bacteria hitherto without such genes, and those which entailed introduction of genes from animal viruses into bacteria." These unfortunately prophetic concerns, coming from within the scientific community, meant that the debate, important as it was, "was largely restricted to technical questions about potential health and environmental risks of the work in question," not the ethics or social consequences of the work itself.[67]

The Asilomar Conference of 1975 accepted the call for a moratorium, but by then researchers had begun to realize that they might find their freedom curtailed, leading the molecular biologists to change their minds as their fear of federal regulation of science overcame their fear of the escape of microorganisms from containment. The biologists argued that "social responsibility need not lead to any erosion of scientific autonomy," and "with hindsight, it is apparent that they managed in the end to keep the discussion on their chosen ground."[68] The moratorium soon fizzled out.

The researchers' fear of regulation, as Fred Buttel puts it, was expressed with "a huge public relations blitz" trumpeting what were said to be the massive productive potentials of biotechnology.[69] The blitz continues unabated, now backed by massive corporate budgets.

Many molecular biologists began to argue that strict regulation would have two consequences: "First, regulation of rDNA research would delay, or even prevent, the development of 'revolutionary' new miracle drugs, vaccines, crop varieties and so on. Second, regulation would cause American R&D to fall behind that of the nation's competitors."[70] This very same rationale was offered by the USDA for its part in developing the Terminator Technology in mid-1998.

The outcome was that the "molecular biology elite" was reasonably successful in heading off mandatory regulation and even more successful, with its exaggerated claims, in attracting the attention of venture capitalists and Wall Street investment houses. "An industry was born virtually overnight," an industry with a great need "for promotion, if not hype, in order to attract venture capital."[71]

&

Today the biotech industry continues, all evidence to the contrary, to make its strident claims that "modern biotechnology" is just "sound science" and the natural evolution of traditional, universal peasant practices. But let us examine those historical processes more carefully. Is the current practice of biotechnology just a new expression of an old tradition, or are there, in fact, radical discontinuities between traditional and current practices?

Chapter Five

Progress it is not

To help feed the world, Monsanto and other biotech giants are tearing up a centuries old farming system and transplanting it in a brave new world where a handful of global corporations, not farmers, set the agenda. —*Christian Science Monitor*, July 30, 1998

A favorite claim of the biotech industry is that what it is engaged in is simply the modernization of ancient and traditional practices of seed, plant, and animal selection and breeding.

A quick review of this natural history, however, exposes the industry claim for what it is, a thoroughly revisionist history, and helps to differentiate the practices of biotechnology, including genetic engineering, from traditional, sustainable practices, as outlined on page 54.

To describe a progression from one set of circumstances and practices to another is not the same as describing this movement as progress. The distinction is fundamental to appreciating the historic novelty of genetic engineering.

Natural or "self" selection

Fluid diversity, or dynamic equilibrium, might best describe the natural order of Creation. Organisms are not precise. They do not stand still, fixed, changeless, and timeless. Organisms cannot be described by a parts list in a computer because any repairman would find quickly enough that the same part could be found in a number of organisms, but performing a different function in each, with that function depending not only on the model itself, but on the environment of the organism. In any case, gene swapping, horizontal gene flow, as well as environmentally and internally induced transformations, are going on all the time, and organisms are constantly evolving and changing to suit the demands of the day and the influences of their lifelines. "The stability of organisms and species is dependent on the entire gamut of dynamic feedback interrelationships extending from the socio-ecological environment to the genes," writes biologist Mae-Wan Ho of Britain's Open University. "Genes and genomes must also adjust and respond, and if necessary, change, in order to maintain the stability of the whole."[72]

In this context one can speak of processes of natural selection, as organisms try out various adaptations, some of which work (they are the ones that fall on fertile ground and multiply) and some of which do not and fall by the wayside, like the seed scattered on stony ground in the Jesus parable.

For adaptations to work, the organisms must be capable of cross-breeding or out-crossing. In plant terms, they must be open-pollinated so the birds, bees, and insects can go from plant to plant, carrying the pollen that fertilizes and creates diversity, or so that the wind can carry the pollen from male flowers to female, even on the same plant.

Because this is also a highly non-directed, promiscuous process, it can also be described as a low-risk strategy of survival. (The biotech industry would probably refer to it as inefficient.) There will be mistakes and failures, but there will be lots of successes as well. And the combination that is a successful cross in one year's cool and wet conditions might be a total failure as far as survival is concerned in the next year's hot and dry conditions. Hence the need for diversity, or "Don't put all your eggs in one basket."

How long this process has been going on depends on your view of how the world began, but we needn't try to solve that problem here. The point is that there is a great deal of fluidity the untrained eye only occasionally notices — like the chive flower without a stem I discovered growing out of the base of a contorted leaf rather than at the tip where the flower normally appears. Were I really a scientist, I would have separated the chunk of it with that strange mutation and let it grow so that I could observe what would happen next.

Human selection

Somewhere in the course of history, humans entered the scene and started to assert their preferences on the flora and fauna they found around them. As long as they were hunters and gatherers, their impact was limited. When our forebears settled down about 10,000 years ago and began to farm, that is, to live in one place and more deliberately produce the food they needed on the land and in the forests surrounding them, they naturally began to select. Certainly some people died along the way, experimenting with new plants or being forced to depend on a single crop when others failed, such as cassava, until they discovered, by trial and error — the science of the time — how to prepare it so that it was nourishing rather than deadly. Such deaths should not, however, be considered tragic or avoidable. They were simply another consequence of being alive in the first place. (The pseudo-science of "risk analysis" is a very recent invention.)

Other choices would have been made on taste and keeping qualities. Since people were no longer moving about, they had to learn to maintain a variety of foods for a variety of weather conditions. They could not afford the high-risk strategy of growing crops that only did well in dry years, or only in wet years. They had to maintain a diversity, and that diversity had to be allowed to continue to evolve.*

Such a low-risk strategy of maintaining maximum diversity and encouraging continuing evolution was essential to the survival of the people. It was a low-risk strategy because that was all they could afford — in the sense that if they had a crop failure, they could not go to the store and buy rice from Vietnam or kiwis from Australia to cover their shortfall.

Vegetative propagation

Closer to our own time, women realized that tubers, bulbs, and runners have different methods of reproduction or replication. Plants could use the familiar practice of flowering and pollination, whether by self or via

---

* I will explore *ex situ* and *in situ* seed conservation, and the concepts behind the practices, in Chapter Eleven.

bees and insects; they could divide, as daffodil bulbs and garlic do; or they could be cut or broken and the pieces replanted, as is the practice with potatoes, to grow into whole plants asexually. Strawberries have another way of reproducing, by sending out runners that take root and form new plants. Thistles and couch (twitch) grass have yet another way of reproducing and colonizing open spaces; they send out rhizomes, roots, which are capable of seizing any opportunity to send up new sprouts that will become new plants.

These forms of vegetative propagation, as it is called, produce clones, but not clones like Dolly the sheep (or Disney's Mickey the Mouse) because while genetically the same, they will also be different as the new plants, tubers, or shoots adapt to and alter their own immediate environments (mutate). Each new plant has its own unique context or environment, like children (even twins) in a family, rather than the clones in a petri dish in a controlled, sterile lab environment.

Hybridization

In this same category of natural evolutionary selection and adaptation combined with human selection, hybridization occurs both naturally and by human intent. Until the 1930s, when F-1 hybrid corn was "discovered," hybridization referred to the process whereby widely differing parents produced offspring that would show "hybrid vigour," or heterosis.[73] The mule, the outcome of the natural breeding of a horse and a donkey, is perhaps the best example of a natural hybrid. While it has greater strength and resistance to disease, as well as a longer life span than either parent, it is also sterile — a natural dead end. Similarly, we could achieve hybrid vigour in our market lambs by using a Suffolk ram on our Leicester-Cheviot cross ewes. The resulting three-way cross would not be a good breeding line, although not sterile, but that did not matter since the purpose was to achieve a good carcass — a dead end as far as the animal was concerned.

Sterility is not the norm for hybrids, though very often the hybrid vigour appears only in the first generation. The process does add to diversity, and farmers have long used crossbreeding to alter the characteristics of crops and animals. With animals, the process of adaptation and, at times, of hybridization has been referred to as domestication. Actually, many farm crops and animals are so domesticated that they would not thrive if turned out into the wild. Domestication creates dependency to one degree or another.

Up to this point, farmers, gardeners, animal breeders, and peasants selected, adapted, and greatly altered the natural order around them. The resulting diversity, however, had nothing to do with "improvement" as the term is used today. It was not "progress." The linearity was absent. Nor did farmers see themselves as breeders of specialized seeds for a global market. They wanted good reliable seed for their own food and that of their neighbors and village. Consequently, seed was traded around and experimented with freely. It was to the advantage of the community to have the best seed possible for the changing conditions of their environment. This was an extension of their low-risk strategy for self-provisioning.

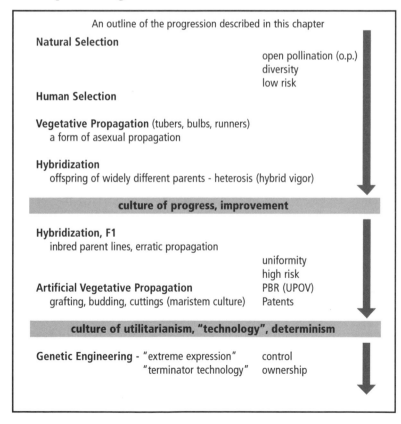

An outline of the progression described in this chapter

Natural Selection

open pollination (o.p.)
diversity
low risk

Human Selection

Vegetative Propagation (tubers, bulbs, runners)
a form of asexual propagation

Hybridization
offspring of widely different parents - heterosis (hybrid vigor)

culture of progress, improvement

Hybridization, F1
inbred parent lines, erratic propagation

uniformity
high risk
Artificial Vegetative Propagation   PBR (UPOV)
grafting, budding, cuttings (maristem culture)   Patents

culture of utilitarianism, "technology", determinism

Genetic Engineering - "extreme expression"   control
                          "terminator technology"   ownership

Radical transition one: The culture of progress

The nineteenth century opened the door to a radical shift in these ancient practices through the global expansion of the imperial powers of Europe and the accompanying ideological imperialism of the Enlightenment and its reductionist science. The schooling systems and the thought patterns of the imperial powers, especially Britain, were replicated in their colonies, and advanced students were brought back to the imperial bases to be culturally assimilated before returning home as teachers and colonial administrators. This pattern is reflected even today in the composition of graduate schools, notably in agriculture, and I saw it with my fellow students at the London School of Economics in the 1960s. Indigenous knowledge, or native science, has had no place in the new scheme of things.

Industrialization (particularly in agriculture, beginning with colonial plantations) and the consolidation of capitalism provided the accompanying structural and economic elements for this historic process. The colonies were used to supply the home countries with cheap staple foods, such as grains, and exotic items, such as pepper and vanilla, not available at home. The local crops were marginalized in the process, much as small farmers around the world continue to be pushed off the best agricultural lands, which are commandeered for the production of crops for export — whether green beans or carnations.

This fundamental economic restructuring was accompanied by the ideology of progress and improvement, which in turn has had a profound impact on the practices of agriculture. The establishment of this culture, with its individualistic, reductionist view of life, its fixation on control, and its dedication to the accumulation of capital as the measure of success and progress, also set the stage for the emergence of biotechnology.

The biotech industry wants us all to believe that its very narrow and peculiarly linear view of the world, like its biotechnology, is universal and eternal, not the product of a very particular culture and historic outlook.

The radical transition that took place within this culture seemed like simple modifications of natural processes, as did my own use of AI and the development of canola that I described in the Introduction. It is, however, precisely the way such incremental changes take place that gives them the potential to be so insidious. The significance of a series of small steps does not become apparent until we have gone far enough to gain some perspective on our journey — or too far to turn back.

F-1 hybrids

Up until very recently, the broadest change in traditional plant breeding was the discovery in the mid-1930s of the F-1 hybrid and its wonderfully commercial function of compelling farmers to buy new seed each year rather than engaging in the traditional practice of selecting and saving seeds from the crops and plants that have performed reliably under local conditions.

This deliberate creation of dependency is one of the chief characteristics of the culture of progress and control, though we are not usually so blunt about it as Daniel Huber, president of Cargill Asia Pacific, who told a World Bank forum on China, "We consider ourselves partners with the Chinese people and the World Bank .... We should not worry whether China can be fed but how best to help China meet her future food requirements .... China needs to choose whether it will adhere to past ideas of food self sufficiency or if it will accelerate its integration into the global food system. The latter choice leads toward expanded imports of bulk commodities such as grains, proteins and edible oils and greater exports of high-value, labour-intensive products like animal protein, fruits, vegetables and fish...A clear embrace of an open food system linked to world markets and based on its own agricultural comparative advantage would be a wise choice for China."[74]

Artificial vegetative reproduction

The second shift was to artificial vegetative propagation, a shift that even now some will view as nothing more than a modest "improvement" on traditional vegetative propagation. Artificial vegetative propagation refers to grafting, budding, and the propagation of cuttings, or meristem culture. This is where the subject enters a gray area, and the line between traditional selection and modern forced intervention is anything but clear. But if we look more closely, it is clear that taking a bud from one tree and grafting it *into* a tree of quite different stock — something that nature cannot do on her own — is not just a quantitative step; it is a qualitative step beyond traditional horticultural practices. While not apparently so, it is in fact an intervention at the molecular level, and thus can be said to be one of the beginnings of biotechnology. This is rather like using *E. coli* to produce chymosin — synthetic rennet — for cheesemaking. In this case the *E. coli* are used to replicate, via fermentation, a synthesized copy of a naturally occurring gene. Is it genetic engineering? A good question — and one that unfortunately never gets answered because the biotech industry insists on polarizing all the issues, marginalizing all the gray areas, which is where the hard judgements have to be made.

A similar question arises with meristem culture, which is the growing of a piece of plant tissue in a nutrient medium in a laboratory. The result is equivalent to a cutting, which becomes a whole plant that is a clone (genetically identical) of the one from which the tissue (or cutting) was taken. Here the primary issue may be that of physical isolation — growing the tissue in an extraordinary and stable environment (not its natural fluid environment, with which it would be evolving) — combined with the ability to produce, in an industrial mode, a virtually unlimited number of genetically identical plants.

This is one point of genesis for monocultural production. It is also another point of beginning for biotechnology, as the identical plants, and the ever smaller pieces of tissue from which they can be grown, down to the level of embryos and cells, make possible the genetic manipulation and experimentation that is the foundation of biotechnology.

What is obviously introduced through these practices is a degree of plant and animal uniformity and, beneath that, genetic uniformity, that is not to be found in the less controlled natural environment. It is obvious to all of us, if we stop to observe and think about it, that nature abhors uniformity. Uniformity is, however, a prerequisite for, and essential to, industrial agriculture. Mechanization and its accompanying "economies of scale," demand uniformity. Big expensive machines — planters, sprayers, harvesters costing a quarter of a million dollars or more — demand that crops be uniform from sprouting to harvesting in timing, stature, and ripening. It's all got to be done in a day, so to speak. A farm machinery operator cannot harvest a single field of corn, tomatoes, or any other crop except all at once. It's an all-or-nothing game.

This is what makes industrial agriculture such a high-risk strategy; it is also why "crop protection" is so crucial. A genetically uniform crop all maturing on the same day is a gilt-edged invitation to insects and pathogens for a grand feast — and an invitation to disaster for the farmer. So the agrotoxin manufacturers and the seed companies logically merge and sell a package deal. The farmer's dependency and vulnerability grow along with the dependency and vulnerability of the seeds and their crops.

What are referred to as "intellectual property rights" (IPRs) are another crucial aspect of this package. IPRs consist of patenting, plant variety protection, plant breeders' rights, and trade secrets. IPRs, in whatever form, are made possible by and require uniformity in the seeds, cuttings, or other type of propagative material, and are deemed essential by the corporations that require a bloated return on their investment in research.

Pioneer Hi-Bred, when it developed the F-1 system of hybridizing corn, had two forms of protection for its property. The first was the fact that the seed from the F-1 crop could be eaten but could not be used with any assurance that it would reproduce uniformly with the same characteristics as its parents. It would not "breed true." The parent lines themselves had been deliberately inbred to produce the required uniformity and to ensure that the resulting cross would be unstable.

The second was secrecy. The company could hold the identity of the parent lines as a trade secret, a legally recognized form of intellectual property protection. This has been used by a wide variety of plant breeders, particularly breeders (companies, that is) working with crops such as wheat, sunflower, and canola that do not lend themselves to the modern form of hybridization (though this is being overcome by determined corporate engineers and their violent technology). The seeds of these crops, not being F-1 hybrids from inbred parent lines, do not carry in them the "protection" of failing to breed true when replanted.

The situation in India indicates what this amounts to in practice. India has a law requiring all importers of germplasm (seeds) to put a sample of the imported seed on deposit in the national seed bank, the National Bureau of Plant Genetic Resources, to ensure that if there turns out to be any problem of disease, invasiveness, or whatever, the identity and source of the seed can be traced. Cargill, when it began its campaign to sell "world class" imported sunflower seed in India, refused to place any samples in the seed bank on the grounds that the Indians would steal some of the seed, begin to reproduce it illegally, and undercut the market for Cargill. Unfortunately, the Indian government let Cargill get away with it.

Modern hybridization has been extended to livestock in the same manner. In fact, Hy-Line International, the world's oldest laying-hen breeding organization, was founded in 1936 as a division of Pioneer Seed Corn Company (now Pioneer Hi-Bred). Hy-Line is the biggest-selling laying hen in the US and is also a strong seller in South America and Asia. As with plants, it is F-1 crosses that are sold, not the parent stock. Hogs are now being made to follow the same pattern of development, with larger livestock and the creations of genetic engineering not far behind.

Ian Wilmut, the director of the Roslin Institute which was responsible for Dolly, the sheep produced from a scrap of tissue from the udder of another sheep (her mother?), described how the big biotech firms in animal breeding "will develop and guard the best genetics, probably allowing access to producers through hybrid animals. Hybrids will contain a mix of genetics, so they will not give away valuable gene stocks. Similarly, major companies are not distributing semen of their original stocks, but rather, producing hybrids they sell to multipliers."[75]

Various legal forms of plant variety protection and plant breeders' rights have also been created to protect the work of breeders of such promiscuous crops. They function very much like a copyright on written material, where the secrets of the construction of the language and thoughts, or information, remain in the head, or head office, of their author. The corporations say that such protection is vital to their work, since they have to be able to guarantee a return to their investors, and to achieve that the ability to patent is absolutely necessary. It is this form of blackmail — "Allow me to patent or I won't work here and you will miss out on progress" — that has coerced governments to adopt repressive intellectual property regimes. I am probably being generous in assuming, in this interpretation, that the governments of the wealthy industrialized countries have not been willing corporate partners.

Radical transition two: Genetic engineering — Smaller pieces, more control

All this sets the stage for the late-twentieth-century emergence of biotechnology and the patenting of life forms as the extreme expression of the culture of domination and control of nature and the ideology of genetic and technological determinism. It is not progress.

On March 3, 1998, the US Patent Office awarded patent number 5,723,765 to Delta & Pine Land Company and the US Department of Agriculture (USDA) for the Terminator Technology described in Chapter One. The patent covers a genetic engineering technique that disables a seed's capacity to germinate when planted ("control of plant gene expression"). It heralds, if allowed to come to pass, the end of agriculture as it has been known for 10,000 years.

Neth Daño of Philippines-based SEARICE described the threat this poses for peasant farmers.

> We work with farmers who may buy a commercial variety, but its breeder wouldn't recognize it five years later. Women select the best seeds every year and over time the rice moulds itself to the farm's own ecosystem. Women also cross the commercial variety with other rice strains to breed their own locally-adapted seeds. The Terminator could put an end to all this and increase crop uniformity and vulnerability. It poses a threat to the culture of seed sharing and exchange that is led primarily by women farmers.[76]

If this technology works and is commercialized, whether the patent holds or not may not really matter. Control of seeds, and thus control of the foundation of the global food supply for humanity, will be in the hands of one or a few large corporations. It will end the ability of the majority of the world's people to feed themselves and will make them dependent on corporate seed suppliers.

Beyond this, if the Terminator Technology is used it could put an end to the food supply for everyone if the crucial genes do not stay put and drift into other plants. The effects could be catastrophic.

Farmageddon, indeed.

Chapter Six

Hormonized milk

We have to convince the consumer that this is good for him.
— Monsanto president Hendrick Verfaillie, 1998[77]

What is it that causes people to believe you can get something for nothing? Or to believe in the promises of the purveyors of a new technology — in this case a miracle drug that forces a cow to give 25 to 40 percent more milk (this was the initial promise) with no harm to itself and for just a little extra feed?

The first genetically altered foods to cause a public stir more than a decade ago were milk produced with recombinant bovine Growth Hormone (the manufacturers of the drug tried to convince us to call it bovine somatotrophin, or simply BST) and the Flavr Savr tomato. No earth-shattering claims were made for either of these products, though the claims for rbGH were a bit absurd and the claims for the tomato somewhat premature. Neither product was described as addressing the problem of hunger or environmental degradation, though it was claimed, without any explanation as to how or why, that rbGH would lower the price of milk.

The products had not been demanded by dairy farmers, tomato growers, or the public, who knew nothing about them until the products had been developed. They offered no health benefits either to human beings or dairy cows, in the case of rbGH, and they met no socially useful purpose. They were simply products of corporate research and development that their proprietors saw as potentially profitable.

That was in the late 1980s, but it was a very long way from where the owners of those genetic "technologies," as they insist on calling them, are today. Then, each technology was the work of a separate company; today they are owned by a single company. Then, they were simply new products for an affluent market, though the lead developer of rbGH had intentions of selling its product to farmers around the world. Today the Flavr Savr tomato is dead — the victim of its developer's arrogance in thinking that all he had to do was tell the tomato "Grow" and it would do so viably on an industrial scale. Mother Nature thought differently of the project. Today, three of the companies initially developing rbGH have dropped out of the game, and the other is marketing its highly questionable product in the United States under the sweet-sounding name of *Posilac.* (If the product is ever approved for use in Canada, it is to be called *Nutrilac.*)

The advertised benefits to the farmer administering the drug to his cows are questionable, but the long-term effects on human health are simply unknown. Since the regulatory apparatus and the science establishment are dedicated to rushing new products to market, including drugs and biotechnology processes and products, the only actual testing of the drug is currently being carried out as an uncontrolled experiment on the American people, who are unknowingly consuming the milk from the drugged cows. They are unknowing because the drug's manufacturer has lobbied, litigated, and intimidated, with near-total success, to make labelling that would indicate whether or not milk comes from rbGH-treated cows virtually illegal.

From the feeble beginnings of those two products ten years ago, to the millions of hectares of transgenic canola, potatoes, soybean, corn, and cotton now being grown in Canada, the US, Argentina, and China, is a long road travelled too quickly. Science, truth, understanding, integrity, gratitude, ecology, respect may all be the hit-and-run victims scattered by the roadside as the drivers of the vehicles of Progress have rushed to make their quarterly payments to their shareholders.*

---

* Production of major transgenic crops in acres:

	1996	1997	1998
RR soybeans, USA	1 million	9 million	20,000 million (40% of total)
Bt Corn, USA		5 million	15-18 million (25% of total)
Bt and/or herbicide tolerant cotton			7 million (50% of total)
Herbicide tolerant canola, Canada		4 million	6.5 million (50% of total)

"We don't need instant success with a magic cow"

Milk, pure and wholesome. A synthesized version of a natural hormone, almost identical to one produced by a cow as a function of its metabolism and lactation. Why would one want to contaminate the first with the second? Monsanto, along with three other companies, is probably wishing it had asked that question many years ago before it decided that developing a drug that would force dairy cows to give more milk would be a really neat — and profitable — business activity.

Instead, Monsanto set about developing recombinant bovine Growth Hormone, using *E. coli* in a fermentation process to produce genetically engineered versions of naturally occurring bGH that were almost, but not quite, identical to that produced naturally by every cow.** Then the company started clinical trials to get the drug licensed for commercial use. The public became the subject of a grand experiment, unfortunately one without the controls necessary to give it scientific validity. The most obvious missing factor is proper identification — i.e., explicit labelling — of the product being tested: milk and dairy products from rbGH-treated cows. If human health problems arise as a result, there is no way positively to identify the cause

Monsanto expected to have its version of the drug on the market for American dairy farmers in 1988. It was February 1994 when it finally made it to market in the US, but it has not been approved and is not on the market in Canada and Europe. The experiment on the American people continues.

❧

In 1986 and 1987 I was thoroughly immersed in a study, sponsored by the Social Sciences and Humanities Research Council of Canada, of how dairy farmers and their organizations, as well as the rest of the dairy industry, regarded the latest technology to be "coming down the road." This time around it was recombinant bovine Growth Hormone (rbGH), recombinant bovine somatotrophin (rBST), or simply BST, depending on the point of view of the speaker. Recombinant bGH is a synthetic analogue of bGH, which is produced naturally by the anterior pituitary gland of animals and humans. RbGH is produced in signif-

---

** Only in Upjohn's version was the amino acid sequence identical to the naturally occurring form of bGH; the others varied by up to 9 amino acids out of 191.

icant quantities by means of bacterial fermentation and extraction — good old *E. coli* at work.

Having been farming for the previous fifteen years, I was intrigued by the fatalism of farmers regarding whatever was defined as technology, and rbGH provided a good case study.[78] It had been pushed into the news in 1985/86 by scientists working for Monsanto at Cornell University, where the initial research and development on the drug began in 1982. Cyanamid had recruited the University of Guelph in Canada to conduct its trials, and there were other trials, some public, some not, like the Ontario dairy herd secretly contracted to Eli Lilly to try out their version of the drug. In every case the milk from the test herds went into the milk supply unidentified.

What I had observed while farming was that virtually everyone involved in agriculture accepted the reductionist assumptions of molecular biology — that we can be reduced to and are determined by the information coded in our genes, and that we have no more control over technology than we do over our genes. At the same time, there is another common ideology (social Darwinism) that holds we are determined by our environment and collective history, so the best we can do is adapt gracefully and without resistance to the world as it presents itself. Unfortunately there will be winners and losers, but the golden rule of life is competition and the "survival of the fittest" according to this social-Darwinist ideology. Those who adopt technology (adapt to its demands, that is) most quickly will be the winners, at least for a fleeting moment, as they ride the "technology treadmill."

Of course such a philosophy induces passivity and a sense of powerlessness that suits those who are intentionally shaping our environment and choices. Impressed by their determinism, we incapacitate ourselves.

It is ironic that dairy farmers were chosen to be the first guinea pigs in this great experiment in biotechnology and social management since they were, at the time at least, very highly organized to run their own affairs. They had the strongest agricultural lobby in the country. They had, and still have, the structures in place to exercise effective control over every aspect of their industry, including permitting or outlawing particular technologies. Yet they talked and acted as if they had no control. There seemed to be no collective inclination to question the purpose of, or social relations embedded in, every technology, nor to question whose technology it was and whose interests it served.

In 1986 rbGH was very widely reported as the first product of biotechnology/genetic engineering that would be available to farmers. Sufficient information was already available to enable an informed, if preliminary, judgement about the drug's effects on cows, on dairy farmers, and on the public if the marketing and adoption of rbGH were allowed to proceed.

The dairy industry could have informed all those involved and made responsible democratic decisions about the application or rejection of this new technology. It did not do so. The industry seemed to be paralyzed. While people involved in every aspect of dairy said they would rather not have rbGH on the market, they virtually all said they had no choice except to use it. The rationalization was, "If we don't adopt it, our neighbor will, and we will not be able to compete."

The larger social context was clearly stated by the more or less official policy bodies:

> "The Science Council of Canada believes that Canadians must grasp the opportunities offered by biotechnology if Canada is to improve its competitive position on world markets."[79]

> According to the University of Guelph, "Rapid scientific and technological progress ... demands increased attention to research and development to guarantee competitive efficiency and thus long term economic prosperity."

> "Efficiency can be maximized by encouraging rapid adoption of these technologies," said Agriculture Canada.[80]

> "Enormous changes are happening in the area of technology ... we cannot afford to be left behind in the race for survival. We all must study the latest inventions and be aware of their implications," agreed the Ontario Milk Marketing Board.[81]

Having been active in farm organizations while farming in Nova Scotia, I knew a lot of farmers personally and was familiar with industry structures. My research project gave me the excuse to visit and interview a variety of dairy farmers. At the time there were about 600 dairy farms in the province, with an average herd size of about forty milking cows.

Farms were disappearing at the rate of ten to twenty per year due to consolidation and increased output per cow.

I talked with my old neighbor, Donny Gunn, who had been farming since 1969 and milking cows since 1977. When I talked with him in 1987 he was milking thirty-six cows in a free stall/milking parlour combination. Donny was very involved in a variety of organizations, so I asked him how much discussion there had been on technology.

"Nothing in the last fifteen years in the egg and pullet association," he replied, "and nothing that I know of in the Milk Producers, though I have been more involved in the board of the [Scotsburn Cooperative] dairy.... We spend practically no time — nil — on technology."

Donny then talked more about his personal outlook. "I guess I am the type that will let quite a few others try new technology. I had the De Laval [milking equipment company] fellow in here pushing me to replace my milking system with the latest technology. They are quite sure, maybe, that it would reduce my milking time by ten minutes and that it certainly improves mastitis problems. But I don't have any mastitis problems ... I have a system that works. It would cost me $3500 to switch my equipment, so why would I do it?"

Not far down the road, Elmer Buchanan shipped the milk he and his wife got from their small herd to Scotsburn. Elmer was a director of the Milk Producers Association of Nova Scotia. His farm had been in the family a long time, and he was not one of the "early adopters" of new technology. He allowed as how he was quite informed about rbGH and feared it more than free trade. "There was a study done in Ontario, and it said 43 percent fewer producers, and 51 percent fewer cows by the year 2000. So if you apply the same figures to Nova Scotia, we would end up with 200 producers in the province .... Our farm — we are milking twenty-two to twenty-four purebreds — would be gone. I figure that anyone under seventy-five cows would be one of the first to go. Quota values will escalate because the larger producers have the cash to go for the quota. They will be the first ones to adopt. It will just be another technology for them."

In another dairying region, one of the largest dairy farmers in the province, milking 185 cows, said of rbGH, "The farther they throw it away the better off we are, because we have too much milk now. If we keep less cows it's going to do harm to the whole industry because you buy less feed, you buy less machinery, you have less labour. Instead of bringing jobs, you are going to lose jobs. It might be good for the individual farm, but for the economy it is bad."

"Do we really need to do with less cows to make the same amount of milk?" asked Alex Forbes, who was milking fifty-five cows. "I would like to see the technology be put to use for gene pools, rare species of plants, environmental problems."

Alfred Scothorn started farming in 1949 with about ten cows. With his sons, he was milking up to 230 cows in 1987. He and his sons were strongly involved in farm organizations, and Alfred's verdict was simple: "There is no discussion, no planning, as far as milk producers are concerned .... There is discussion only when there is a crisis."

Floyd Cock farms with his wife Cheri near Scotsburn. Floyd grew up on a nearby dairy farm. Viewed from the end of the long lane, theirs is a picture-book farm. Their cows are tied up in an old but attractive and modestly renovated barn, and two people do the milking. Apart from the computer (used for bookkeeping) in the house, the only technology is mechanical: silage handling and manure handling. "Other than that, it is a hands-on operation." With forty-five cows milking, the Cock farm is right on the provincial average, but the averageness stops there. Floyd Cock's language is different: he seldom, if ever, uses the words "efficiency" or "productivity." Instead he speaks about "stress" and "comfort."

> Well, I've surpassed all my goals of productivity I've set over the years...I did set a goal about three years ago to go to 20,000 pounds per cow [per 300-day lactation], but I can't do that without hurting the cows, without stressing them more. I'm limited because of the size of the barn and the size of the stalls, and I need larger animals to get that kind of production...If you want to know my deepest thoughts, I can see this place twenty years down the road with 300 animals cooperatively owned by as many as six farmers, all sharing the same milking facilities, each one with thirty to one hundred cows, individually owned and managed herds, allowing the operation to have continuity from generation to generation or from partner to partner. If a person likes dairying, likes animals, and likes working with animals, the facility here could buy and sell to people coming in. A very modern parlour system,

costing $200,000 to $250,000, should be shared by more than one farmer.

My knee jerk reaction [to rbGH] is, why do we need it? I can't answer that one yet. When I first heard about it, I did think it would be nice to milk thirty cows instead of forty-five, but thirty cows under stress — and I think it is going to be stress related — because genetically there are cows now that can do better than somatotropin can. We can breed cows for high production without a lot of stress, and I see a lot of progress in that area already. We don't need instant success with a magic cow. It isn't necessary, we don't have the markets, and it would create unemployment.

I asked Floyd if his opinion was shared by other farmers. "I hope it is," he responded, "but I suspect that it is not because of one very important difference: I like milking cows. I like not milking them, by times. But I do like working with animals, and I consider myself a dairy herdsman. I'm afraid a lot of dairymen, who own dairy farms, don't share this attitude of a love of the cow that I have. They would really rather see less animals in the barn, less forage to make, less labour to be hired to look after the animals, less calves to deliver. I do have quite a nice efficient farm right now with usually two or three workers on it, including myself, and I don't see any benefits. I don't have a desire to milk less animals."

To find out what was being taught to future dairy farmers, I talked with Alan Fredeen, who was teaching ruminant animal nutrition at the Nova Scotia Agricultural College. He was worried that some of the research money came through industry, and government was encouraging that more and more. "We are not questioning it. We're pushing it as being a good thing, and I have been guilty of that in the past. I think we have got to realize that we have spent — the taxpayer and private industry — a lot of money on so-called biotechnologies, and rbGH is going to be the test case."

I asked Fredeen what orientation the students got towards technology. "Well, in the past I have been guilty of pushing higher and higher levels of technology, but in the last year or so I have come to realize that to a large extent we are pawns of industry, Even the way in which I was taught, myself, we never questioned the technology. We accepted that's what was coming and that we had better get used to it as soon as possible."

The major dairy processor in Nova Scotia is Scotsburn Cooperative. James McConnell, its president, sat on the Nova Scotia Dairy Commission, the regulatory/management agency for the dairy industry in the province, on behalf of the processors. He took the lead in formulating the position of the Dairy Commission on rbGH. I asked him how much education had been done by Scotsburn for its shippers. "I would say we are hiding it as much as we can, we're trying not to even think about it. I'm not saying it's right, but that's what we are probably doing...What we don't want is public discussion, like the Consumers Association becoming involved in the thing, and that's why we have not put out any newsletters, it's not come up at any of our zone meetings...it's just not an issue right now."

Roger Mason, secretary of the Nova Scotia Dairy Commission, told me that the work of the Dairy Commission tended to be "to provide a mechanism for regulations either for or against whatever happens to be coming along."

As for the organization that is supposed to be the national voice of dairy farmers, Dairy Farmers of Canada (DFC), their January 1987 policy statement included this paragraph:

> #43. The Canadian dairy industry must keep abreast of new technology in order to ensure that it remains competitive. An example of such new technology is the possible use of somatotropin which could potentially result in significant increases in milk production per cow. The prospect of lower costs of production which may result from this technology could benefit both producers and consumers. Before such technology is adopted at the farm level, however, a complete analysis must be made of its safety, long term effects, impact on the structure of the industry and on genetic evaluation programs, and the consumer reaction to its utilization in the food chain.

There is no mention of farmers or people or animals; only functions such as producers, consumers, and "production per cow."

DFC published a background paper a few months later that failed to raise a single question about the value of rbGH to the dairy industry or its potential impact on cows, dairy farmers, or consumers of dairy products. It expressed a simple fatalism about "technologies": "Along with other technologies aimed at increasing the milk production per cow, the introduction of somatotropin will probably accelerate an already irreversible tendency towards fewer farms and fewer cows."[82]

The language of the paper's conclusion considerably strengthens this determinist orientation and provides some explanation as to why dairy farmers see little or no choice as to what will happen:

> As with any new technology, it will take a few more years before the impacts of somatotropin are fully assessed. Some questions still need to be answered before dairy producers can finalize their position on the introduction of this new technology. For example, what will be the actual rate of response on commercial farms, the long-term effects on health and repro-duction of the cow, the mode of administration ...and, above all, the consumer reaction... Before somatotropin becomes widely adopted, there will have to be surveys on consumer attitude towards the product and most probably some form of educational campaign will have to be initiated...
>
> Wide adoption of somatotropin will probably accelerate the present tendency towards a rationalization of the industry .... Can we stop this trend? It is seldom that progress has been halted .... If somatotropin becomes available, producers will use it as they did with other technologies.

The only dairy organization that advised strongly against the adoption of rbGH — on grounds that it could only hurt the marketing of dairy products — was the Canadian Dairy Council, which represents the dairy processors!

"Lack of catastrophic health effects"

RbGH is certainly not the first "technology" to affect the dairy industry, but it provides an opportunity to look at the culture that produced it and that it, in turn, nurtured.

Since the 1950s the dairy industry in North America has been trans-formed by the adoption of a number of significant technologies. On the farm, the milking machine and bulk milk tanks were the earliest of the mechanical technologies to hit dairy farming. These were followed by the pipeline milker and the milking parlour. The newest technology in the barn and milk house is the computer. It is used not only for manual record-keeping, but also for automatically recording milk yield and for dispensing feed. These latter operations are made possible by means of a transponder hanging on the cow's neck. that electronically identifies the cow in the milking parlour and at the automated feeder.

Apart from the widely influential developments in crop production and animal nutrition, milk production per cow has received its greatest boost from continued genetic improvement through selective breeding and then artificial insemination. The latest "technology" in this line is embryo transplant and accompanying genetic manipulation. The aim of it all is to increase the "productivity" of the dairy cow, usually with the expense of a short life span and high maintenance costs.

In the literature of the dairy industry, little mention has been made of specific technological innovations. It would appear that few or no records have been kept, and little or no significance attached to the possible or actual results of technological innovation. The history of technology simply does not seem to have been of interest.

However, as a result of the large amount of publicity generated in support of rbGH by its developers, these same developers have said over and over that no new technology had ever been so thoroughly discussed prior to its general availability.

BGH had been known about for years, but it was the discovery of how to use genetic engineering techniques to produce it synthetically that opened the door to the large-scale development of the drug in the 1980s.

As already indicated, the first clinical trial of rbGH was conducted by Dale Bauman at Cornell University from 1982 to 1985, under contract to Monsanto Company. The results of this single three-lactation trial with about thirty cows were published at the end of 1985, along with an analysis of their economic significance carried out by Cornell agricultural economist Robert Kalter. Virtually all subsequent public discussion has started from this incredibly limited data base. Bauman and Kalter claimed an increase in milk output "by up to 40 percent" with no observable detrimental effects on the cows that could be attributed to the rbGH injections.

Kalter and Bauman assumed the roles of leading advocates — saviors, to hear them speak — in any discussion of rbGH, and Bauman continues to loudly defend and explain the drug while trying also to silence critics.

> If the promise of the "new biotechnology" is fulfilled, the benefits to society are obviously greater economic efficiency and an improved standard of living. On the other hand, the speed with which new biotech-related products or processes are commercialized will impact established methods of conducting the world's economy with resultant dislocations, equity impacts and alterations in social structure.

> ...Increases in production imply a reduction in consumer prices [and] declining national dairy farm numbers...In the longer term prices must decline, accelerating the withdrawal of farms from the sector .... Clearly the short term impacts of the rapid adoption of bGH could be harsh while a new equilibrium is reached .... Should bGH become widely used and prices allowed to adjust, it is unlikely that nonadopters could survive.[83]

Bauman defended the integrity of his work, while acknowledging the support of various chemical and drug companies, by pointing to "over 60 papers and abstracts on the effects of exogenous somatotrophin" that he and his research group at Cornell had published. At the same time he complained that with all the media notoriety his work had received, three points were frequently erroneously stated about his efforts. "To set the record straight," he said,

> 1. We have never conducted a proprietary study [a study for private interests] in which the data wasn't published. As soon as we get a study analyzed and summarized, we publish it!

> 2. We have never sent sick cows to any company or clinic anywhere in the world. We have simply never observed somatotropin treatment to cause any sickness!

> 3. We have not been selective in our published data by excluding data from sick cows. To do so without addressing this in a publication would be unethical and unacceptable![84]

Bauman explained the more ambiguous results obtained by trials at other universities as the result of stress, not the effect of the drug.

"Stress is difficult to quantify, but stressed animals would expend greater than normal energy (heat) for maintenance, produce less milk, and have a lower productive efficiency (milk/feed). None of these stress related effects have been observed in cows treated with somatotropin .... In contrast to the clear lack of catastrophic health effects with somatatropin treatment, there are not sufficient published data to allow for an evaluation of subtle health effects .... However, our study involved only 30 animals in a well-managed herd."[85]

Bauman's use of the term "catastrophic" is striking, as is his simple statement that his database is rather limited — only thirty animals! Apparently the fact that no cows dropped dead meets the criterion of "lack of catastrophic health effects." The term continues to appear in discussions of biotechnology as its advocates point to the absence of "catastrophic effects" as proof of safety (see Chapter Nine on regulation and burden of proof).

It is also an indication of how biotechnology has been approached as a science when Bauman arbitrarily assigns any symptoms of stress — from laminitis (shedding of hoofs) to mastitis — to higher milk yield rather than to the use of rbGH, even though the higher milk yield and accompanying stress is forced upon the cow by the administration of the drug.

Up to 1988, the primary focus of all the bGH trials was efficacy: does it work, and how well? Bauman concluded in his 1985 report that, on the basis of "the available information," there were "no adverse effects" from the use of bGH. It later emerged that when Bauman referred to "the available information" he was not lying, he was simply referring to the information that Monsanto had made available to him. A Cornell masters student, Tess Hooks, discovered that all the raw data from the trials had been transmitted by a technician directly to Monsanto via a modem in the dairy barn. Monsanto, in turn, sent the data on efficacy to Bauman. It did not send back whatever data there might have been on side effects and animal health, perhaps because the trials were being conducted in order to gain regulatory approval for the marketing of a drug, and the issue for Monsanto was efficacy, not the health effects on cows or people.

Tess Hooks had chosen to do her masters thesis under Fred Buttel on the subject of industry-university relations as exemplified by the contractual and working relations between Cornell and Monsanto. Hooks documented the company's control of data and its analysis by going right to the Cornell-Monsanto contract, which included the following clause:

> 4. REPORTS Data generated in the INVESTIGATION shall be promptly reported by INVESTIGATOR [Bauman] on a continuing basis to SPONSOR [Monsanto] for compilation and analysis. The summarizations compiled by SPONSOR will be promptly reported to INVESTIGATOR.

She observed, "The implications of this seemingly innocuous paragraph are ominous for open scientific discussion. Consider the fact that even the principal investigators may not have unimpeded access to the data generated by the research. That is, the principal investigator must not only rely on Monsanto's willingness to share the information; the investigator must also be willing to accept Monsanto's statistical summaries and analyses. In addition, since Monsanto effectively owns these data there is no guarantee that they will become available even after the publication restrictions on the investigators has expired."[86]

In 1995 Hooks, Fred Buttel, and David Kronfeld[87] submitted a manuscript, "Scientific Conduct: The Mischaracterization of Mastitis Associated with Recombinant Bovine Somatotropin," to the journal *Science*. The article was rejected, and a careful reading leaves little doubt as to why. The paper begins:

> Inaccurate or selective presentation of data is a form of scientific conduct that may be subtle and escape attention. It may reflect contract-constrained communications and conflicts between the interests of a company and the public ....

> In 1994, the relevant mastitis data [covering trials that commenced in 1984] were released, albeit obscurely, and it is now clear that the principal investigator mischaracterized his results not only in the *Journal of Dairy Science* but also to the U.S. Senate's Office of Technology Assessment and to the White House ....

> Mastitis associated with metrbST administration has been characterized by persons under contract to Monsanto in three ways: as postulated but never observed; as referable to

increased milk production; and as minor and manageable. All three characterizations can be shown to be incorrect ....

Monsanto's pooled presentation flouts several fundamentals of epidemiology and preventive medicine. No criteria were given for the selection of 15 studies from at least 30 published Monsanto studies or for the exclusion of the other 15 studies from this alleged "entire data set" ....

... The principal investigator, who is a Cornell professor and Monsanto consultant, drafted the biological section of a politically important document on rbST prepared for the U.S. Congress in 1991 which stated that "[c]atastrophic effects such as the incidence of ... mastitis ... suffering and death have been postulated to occur. However, no such effects have been observed with bST supplementation of dairy cows in any scientifically valid published studies, nor have subtler health effects been in evidence." This statement fails to discriminate between observing and reporting. *

Hooks, Buttel, and Kronfeld conclude: "If Monsanto truly believes that science is a progressive force in society and that the scientific method is a self-correcting process that ultimately produces the 'truth,' it should make its health data available for overdue independent evaluation."

In a 1988 editorial, the reliable journal *New Scientist* noted that Monsanto had begun "a propaganda campaign to convince the public that BST milk is no different from ordinary milk." The editorial pointed out that Monsanto had issued a press release describing trials in Europe and North America and explaining, "The purpose is to confirm that BST is safe and effective in improving the productivity and efficiency of milk production." The editorial concluded, "Presumably scientific trials that do not confirm BST's safety and efficacy are not scientific." The observation remains appropriate.[88]

&

Trials with rbGH were conducted for the American drug and chemical company Cyanamid at the University of Guelph, Ontario, from 1995 to 1998. Dennis Lawson, then director of corporate development for

---

* Voluminous references have been deleted for the sake of readability

Cyanamid Canada, told me personally about the Guelph trials.

> There is absolutely no indication that there is anything detri-
> mental to anybody, as we would expect, because this is essen-
> tially a naturally occurring chemical.

> The implication of the use of this product is that it is basically
> making the cow more efficient .... To my knowledge there is
> certainly nothing to indicate that the cow is being pushed
> harder. They are just having to work less, with less.

A week earlier I had visited the Elora Research Station, where the trials
were being conducted, and talked with one of the herdsmen about the
condition of the cows and the consequences of using the drug. I had
mentioned this to Lawson before our conversation began. The herdsman's first-hand account was not quite what Lawson told me.

> Some cows just can't take it ... it's stress again, you're creating
> a highly stressful situation here, with feeding, and producing
> more than is naturally produced .... It's taken its toll, it's wiped
> out a few...We shipped one here a week ago and she was just
> skin and bones. She was eating pretty well what was put in
> front of her, and putting out the milk, but I guess it took more
> than she could get. Her body just deteriorated, she had a feet
> and leg problem as well .... No udder problems...though some
> of the heifers were producing so heavily that when it came
> around to the second lactation they blew their udder to pieces
> ... a lot of twins, trouble breeding them back .... We are looking
> at long-term effects, but unfortunately I am under contract
> and cannot tell you anything other than what is in this paper
> [the *Dairy Research Report*] in terms of effects. It means that I
> have to give prior information to the corporation that pays for
> the work.

This sounded rather like what was going on at Cornell.

Manipulating genes, language, and regulators

On September 20, 1985, Judith Juskevich of the US Food and Drug
Administration (FDA) issued a memorandum that allowed Monsanto
"zero withdrawal time and zero discard period" for cows treated with
daily injections of rbGH. This meant that milk from cows on trials and

treated with rbGH was being sold to the dairy processors and consumed by the public without their knowledge or consent.

The years after 1985 were years of controversy, and in spite of pro-
fessional management for corporate image and science reporting, by
the late 1980s the drug companies felt an urgent need to engage in
damage control. Dairy industry representatives and drug company
executives met and agreed that it would be very helpful to have an article appear in a reputable, refereed scientific journal explaining how
sound all the science, and the approval process, had been. They needed a whitewash, in other words.

This is how it came about.

"The four companies developing BST, American Cyanamid, Elanco,
Monsanto and Upjohn, are AHI members. These four companies
formed the AHI BST Public Information Working Group in late 1986 to
address a public relations challenge of mutual concern," wrote Steve
Berchem, public information manager of the Animal Health Institute
(AHI), in a letter to Richard Weiss of the National Dairy Promotion and
Research Board in March 1990.[89]

A year earlier, the AHI BST Working Group and representatives of
the National Milk Producers Federation (NMPF) had met and decided
that the "NMPF will seek an independent, third party review of the scientific and economic impact of BST, perhaps by the National Academy
of Sciences. If necessary, NMPF will also ask the Food and Drug
Administration (FDA) to delay commercial approval of BST until the
review is completed."[90]

A few months later, "Mr Barr [ceo, NMPF] reported on a conversation he had with Dr. Gerald Guest, [director] FDA/CVM [Center for
Veterinary Medicine], regarding CVM's plans to prepare a food safety
'white paper' about BST...FDA's willingness to prepare such a white
paper is an unprecedented move by the agency."[91]

The outcome was a lengthy feature article in *Science* magazine
(August 1990) reviewing the FDA approval process for rbGH. The lead
author of the article, Judith Juskevich, was not identified as the person
whose work the article was reviewing.[92]

The article stated that FDA scientists "summarized more than 120
studies that examined the human safety of recombinant BST" and
came to the conclusion that BST poses no risk to human health. It also

reported on an FDA-sponsored joint advisory committee meeting in May 1990 that dealt with the labelling of food products derived from BST-treated cows. "Based on the conclusions of the committee members, and on its review of the facts, FDA has concluded that it lacks a basis under the statute to require special labeling of these foods. Food companies, however, may voluntarily label their products provided the information is truthful and not misleading .... " And the article quoted FDA Commissioner David Kessler: "We have looked carefully at every single question raised and we are confident this product is safe for consumers, for cows and for the environment."

I had the opportunity to interview Judith Juskevich in March 1991. She told me that she had been (and still was) a consultant to the FDA since 1982, the year Dale Bauman started his work on rbGH for Monsanto at Cornell, and that she was the FDA official who gave Monsanto approval to sell the milk from its test herds for public consumption. Her article in *Science* was peer-reviewed by Dale Bauman, among others, but Ms. Juskevich told me that she did not know that Bauman was one of her reviewers.[93] I do not know who asked Ms. Juskevich to write the article, but the careful description she gave me of the FDA's new drug approval review process is worth noting in detail.

> When a new drug comes in, they get an "investigation of a new animal drug" application. For the investigational studies they do for efficacy, I'd say with most drugs for most animals it is almost not feasible for them to do those studies unless at some point they can start to market either the milk or the meat from the animals because of the large number of animals used in those studies. And you know, to tell someone to burn or bury a thousand head of cattle, or pigs, seems ludicrous if the meat and milk is actually safe ....

> The companies send in interim food safety information along with doing the rest of their studies for efficacy and animal safety and environmental issues. There is a set of guidelines at FDA in terms of what they need to do to show that the products will be safe for human consumption.

> So they can get an investigational withdrawal period prior to submitting all their information, and those are generally very conservative estimates of how long the animal has to be off

the drug before it goes to market. If they complete all the studies that they need for human food safety, all the toxicology studies and all the rest of the chemistry studies, and they are considered adequate, then they can get whatever withdrawal period they are going to get when they actually market the product. So the product does not have to be actually approved for them to get a withdrawal period or a milk discard period ....

I asked Ms Juskevich if it was the case that there was already no withdrawal period for rbGH in 1985. She responded, "Yes, they had submitted all the studies that we requested .... The FDA requested — the studies that FDA requested of all of the companies, well actually I'm not sure all the companies have actually finished all of them — well, but they did all of the studies that they needed to get the zero withdrawal. So the company actually gets that withdrawal period for their investigational purposes."

I told Ms. Juskevich that I understood one of the requirements for licensing was that there had to be an adequate and reasonable test for residues.

FDA does not require a testing method for any drug that has a zero withdrawal period ....The issue of residue testing of the growth hormone itself was considered totally irrelevant for several reasons, the main one being that bovine growth hormone is *not* active in humans, even if you inject it. So in a sense it doesn't really make any difference if it is increased or not because it will never have any effect anyway.

I could hardly believe my ears. As she described it, the whole approval process was based on circular logic and conjecture!

After I caught my breath, I asked about Insulin-like Growth Factor 1 (IGF-1). I knew this was a factor that the FDA had ignored. Writing in the MIT journal *Technology Review*, Wade Roush conveyed the concerns of one of the most consistent and respected critics who had questioned rbGH and FDA approval of the drug:

RbGH acts on lactating cells in the mammary gland through a messenger substance called insulin-like growth factor1 (IGF1). The same chemical acts as a messenger for human growth hormone in children and adults. Milk from rbGH treated cows has been found to contain higher than normal levels of IGFI, which may survive digestion and enter the human blood stream. These extra doses could cause premature growth in infants, excessive development of the mammary glands in male children, and breast cancer in women, according to Samuel Epstein, a professor of environmental and occupational medicine at the University of Illinois Medical Center.[94]

In response to my questioning, Judith Juskevich said:

> We had a lot of information about IGF, normal levels in humans and the likelihood of it being absorbed and of being orally active, and at that time we actually did not think it was a major issue ... so we allowed the marketing of the milk to continue but it was finally decided to ask the companies for the studies on it anyway and basically all the studies actually showed pretty much what we figured they would.

Roush also pointed out that, in the August 1990 issue of Science, "FDA researchers Judith Juskevich and Greg Guyer concluded that milk from rbGH treated cows did show IGFI levels at least 25 percent above average. However, they maintained that studies in rats show the substance is broken down in the digestive tract, and therefore could not cause any important physiological effects."[95]

I asked Ms Juskevich about the article in *Science.*

> We ended up writing the paper I guess because we had been asked for it — they had asked for a National Academy of Sciences review. But I don't really feel that at least the people that I knew that were involved felt like they really wanted to be in the position of supporting the product for the company. It was difficult to even talk about it with looking like you did.

At that point I said that I was beginning to see that one has to look at the assumptions of how science is done. I suggested that there was something in the assumptions of the FDA, or Health and Welfare in Canada, that really needs to be questioned because there had been a lot of blatant promotion by Monsanto and a lot of other agencies.

> Well, this was a very, very unusual circumstance, at least for a veterinary drug...I think generally you are just not allowed to talk about anything that comes in. You are never put in a position where that is an issue because everything that comes in is

confidential and you are not even allowed basically, legally, to even say that you have any particular drug from a company in for consideration. So I don't think people have been put in that position before. FDA just never discussed any drug.

Actually, I don't even know how people found out they were working on it, on bovine Growth Hormone. I can't remember what actually started it. But once it started to get publicized, the company said you can release the human food safety information. And then we were allowed to actually talk about the human food safety study.

What about animal health and the effects of the drug? I asked. Didn't that enter into consideration?

For marketing? No. They are totally separate issues. In terms of approving the drug for commercial use, they will have to finish everything before it's actually on the market, but in terms of the meat or milk from treated animals to be marketed prior to approval, that only has to do with determining the human food safety of that product .... It's handled by different groups. The office of new animal drug evaluation is divided into different divisions, and there were two divisions that were concerned with human food safety and those divisions worked very closely together. But there were other divisions that were concerned with efficacy or target animals, so they worked essentially separately. There was interaction, but in terms of how a company proceeded along with the different phases of it, people in human food safety were not really concerned with efficacy or the target animals. But to get approval, they have to finish the human food safety, the efficacy, the target animal safety, and the environmental concerns for the product.

... So it seemed like FDA was being criticized for sort of supporting this drug that might be economically disastrous for dairy farmers, but really all people were doing was saying these are the studies, this is what we asked for, this is what we got, they were well done, and the studies show it safe for humans and there are no safety problems, or for efficacy studies, they show that it works ... and once they have done all those studies, the FDA approves the product and it really has no license to say we really think this is not a good product and it should not be out there.

Judith Juskevich may well be sincere and believe what she told me. What she told me, however, provides a disturbing picture of reductionist science in practice. There are, both structurally and philosophically, a number of disparate pieces, but no whole.

&

In May 1993 the US Department of Health and Human Services announced that the FDA had approved "the new animal drug sometribove" to be marketed under the trade name Posilac. In the announcement, FDA Commissioner David A. Kessler stated, "This has been one of the most extensively studied animal drug products to be reviewed by the agency. The public can be confident that milk and meat from BST treated cows is safe to consume."

The issue of conflict of interest, however, keeps cropping up, as should be obvious by now. A 1994 report from the US General Accounting Office (GAO), under the heading of "ethics," reports on its investigation of conflict of interest charges levelled at three FDA employees who were in some way involved with Monsanto.

In the first case, the GAO's conclusion was that there was minor rulebreaking by Dr. Margaret Miller, who had worked with Monsanto and concluded there was no test available to detect the presence of rbGH in milk. She continued to publish on behalf of Monsanto while working at the FDA.

The second case was that of Michael Taylor, then head of the USDA's Food Safety and Inspection Service (FSIS), who had been an attorney with a law firm that represented Monsanto. In that position he argued that there was no need to label rbGH milk. Subsequently at the FDA, Taylor worked on labelling guidelines that concluded virtually the same thing. The GAO said this was not an ethics violation.

Susan Sechen, the third case, had been a graduate student at Cornell, assisting with Monsanto funded studies. She was later the lead reviewer of scientific data on rbGH for the FDA. The GAO concluded that because Sechen was not paid by Monsanto for her work as a graduate student, there was no ethics violation.[96]

If some were favored, others were not. Richard Burroughs, the veterinarian in charge of its rbGH review from 1985 to 1988, was fired by the FDA in November 1989, allegedly for incompetence. "Burroughs says he was pushed out because he pointed to important flaws in the companies' safety studies that his superiors had overlooked in their eagerness to approve rbGH. 'It used to be that we had a review process at the Food and Drug Administration. Now we have an approval process,' he said last year."[97]

Unfortunately the case of rbGH does not seem to be unique. There is mounting evidence that scientific and peer-review journals abuse their esteemed position to cater to special interests. An editorial disputing the link between high blood pressure and salt consumption in *Science*[98] was signed by David McCarron. McCarron is a paid consultant to the Salt Institute, representing the salt industry, but that was not pointed out in the editorial.

The *New England Journal of Medicine* published a review of Sandra Steingraber's *Living Downstream: An Ecologist Looks at Cancer and the Environment*, in which the reviewer, Jerry H. Berke, MD, accused Steingraber of producing a biased work because she argues that cancer is a social-environmental illness far more than a "personal life-style" disease. There are many questions about the validity of Berke's argument and the evidence he cites, but what is most shocking is that the journal did not identify him as director of medicine and toxicology for W.R. Grace & Co., one of the largest chemical companies to be found guilty of numerous deleterious environmental and human health practices.[99]

The *Journal of the American Medical Association* published a review of 106 published scientific papers disputing the claimed links between secondhand tobacco smoke and lung cancer. The review concluded that "the only factor associated with concluding that passive smoking is not harmful was whether the author was affiliated with the tobacco industry."[100]

This is the culture of biotechnology.

Give me my blinders

Looking back over newspaper accounts, research reports, and journal articles from the 1980s on rbGH, it is hard to believe that scientists, dairy farmers, farm organizations, and corporate representatives could all be so eager to embrace and endorse, without thought or evaluation, this new "technology." Little, if any, consideration seems to have been given to the cow, the object of exploitation, by anyone other than a few dairy farmers and anonymous university herdsmen and women. No consideration was ever given to public sensibilities about added hormones in the food system, or the unknowns such as IGF-1.

This blind commitment to new technologies should have caused many scientists to rebel at the prostitution of their discipline, but the attitudes expressed in two research reports from the province of Alberta's ministry of agriculture, for example, indicate all too clearly why scientists went along with the hype: research money, career advancement, and unquestioning identification with the dominant culture.

"University of Alberta [U of A] Animal Sciences researchers Dr. John Kennelly and Dr. Gerry de Boer are hoping their project...will reveal milk production increases of 15%-20%. And, says Dr. de Boer, a 40% increase with a well-managed herd is entirely possible."[101] It is striking that the research results were announced before the research had even begun.

Two years later the U of A reported that studies on rbGH were underway at Oyster River Farm on Vancouver Island. The report is unusual in its acknowledgement of how little the scientists understand. "BST is a kind of 'chief director' of the cow's endocrine system. For reasons not yet understood, injecting a cow with BST instructs the cow's system to give top priority to producing milk, rather than making her grow bigger, or store more fat." The article talks vaguely about lactose, glucose, and insulin and then says, "There may be other chemicals released into the bloodstream, but if so, scientists haven't discovered them yet. This all may sound very complicated, and it is. The insulin-like growth factor #1, or IGF-1, is a bit of a mystery itself."[102]

In the same report, U of A food scientist Lech Ozimek suggests that "more thorough milk analysis is needed to see whether any other organic compounds such as the hormone IGF-1 are 'leaking' into the milk of BST-treated cows .... Many other organic compounds may be involved ... and some of those compounds may end up in the milk. So far nobody is looking for them." Ozimek, according to the article, sought funding to do the research that was obviously needed, but he was turned down by the companies making rbGH and by the Alberta government.

Three or four years after the clinical studies had begun at Guelph, during which time the milk from the test herds was going into the dairy pool without the knowledge of either farmers or the public, Brian McBride told me, "When we complete this work, I would like to be able to say, one way or the other, that yes, this milk has or has not these materials in it, it is safe or unsafe .... Somatamedins, or IGF...are naturally secreted into milk, but we want to know if there are any more secreted due to BST treatment, and whether, in fact, if they are secreted, they have any effect."[103]

It appears that neither McBride nor any other dairy researchers ever got the answer to the questions, whether through loss of interest or, more likely, loss of funding. Corporations such as Monsanto are not likely to pay for research into subjects about which they wish to remain ignorant.

&

The attitude of contempt for the public that seems to be a plague of the biotech industry is exhibited in an article in the *Journal of Agricultural Ethics* by Jeanne Burton and Brian McBride.[104] Dismissing public questioning of the need for and use of rbGH as "hysteria," the authors state that, "Attention to the concerns of the public may be the only means to prevent hysteria over this and future products of biotechnology ... Misinformation, from unqualified (and often unidentified) sources pours into the media. This misinformation causes undue hysteria among dairy producers and consumers." Burton and McBride conclude: "It concerns us that the public hysteria leading to the ban on the sale of milk from rbST-treated cows in clinical trials across Canada has effectively brought a halt to much of the industry-funded research in this country...The pursuit of understanding is the essence of science."

Unfortunately, the pursuit of understanding seems to have taken a backseat to the pursuit of new products and corporate research grants. This is illustrated by the intentional manipulation of language and meaning, along with the manipulation of genetic "information," that has been engaged in to confuse, if not deceive, the public as well as farmers. The tug-of-war between biotech promoters and public critics over the name of recombinant bovine Growth Hormone is a good example. The industry has tried to make the name rBST (or simply BST) stick, but critics have insisted on using the original name of rbGH.

Until the end of 1986, the synthesized analogue of the cow's pituitary gland hormone controlling growth was universally referred to as rbGH. Since then the industry has assiduously referred to it as somatotropin. I asked a leading animal science researcher, who was conducting trials with rbGH at the time, why he used the term bGH in his published reports. He explained that he used the term when he referred to earlier research, though the current term was "somatotropin." I asked if the switch to "somatotropin" at the end of 1986 was accidental.

"No," he said, "it sterilizes 'hormone' .... It sterilizes the connotation. Bovine Growth Hormone is the correct term .... There are a lot of farmers who might use a product in the future under one or another name, and they might not even realize, unless they are properly educated, that they are using a hormone. It is very different than a steroid hormone, and I think that's probably why the companies wanted to sterilize it."[105]

Shortly before that conversation, in 1987, the trade journal *Dairy Foods* had published an explanation: "Five leading agricultural product companies have developed a program that is designed to educate producers, processors, and eventually consumers about bovine growth hormone. Education begins with the product's name. Bovine growth hormone...is more accurately called bovine somatotropin (BST). The product's 'hormone' label, although accurate, is one of the major stumbling blocks its proponents face. 'It is a hormone ... but people are sensitive about that word, especially when it affects a fundamental food. So we need to carry out an information program,' says Laurence O'Neill, manager of public relations for Monsanto Agricultural Co."[106]

Posilac bovine somatotropin was approved for commercial sale in the US by the FDA in November 1993, and commercial sales of *Posilac* began February 4, 1994, according to Monsanto's annual report for 1993. A Protiva (Monsanto) Status Update sheet in May 1996 reported that since its introduction in 1994, approximately 15 percent of all dairy producers in the US had purchased the product. Monsanto's figures, not surprisingly, reveal very little about the actual usage of *Posilac*. A few months later a USDA survey found that only 10 percent of the nation's dairy cows were receiving rbGH, and Monsanto did not dispute the USDA's reports of its sales volumes.[107] Reports in the summer of 1998 indicated that rbGH usage levels in the key dairy states of New York and Wisconsin was 4 percent or less, though usage rates may be higher in California and Florida, with their industrial herds of 1000 to 3000 cows.

Given the findings of recent research into IGF-1 drugs on the market or in trials for human use, we can be thankful that more farmers have not taken to injecting their cows with *Posilac* and that it is licensed for use in only a few other countries. Elevated levels of IGF-1 are now showing an alarming correlation with increased incidences of prostate and breast cancer. Independent scientists, such as Samuel Epstein, cited earlier, have been warning about this for a decade.

One study, published in May 1998 in the British medical journal *Lancet,* commented that while IGF-1 is a naturally produced hormone that is necessary for normal cell growth, elevated levels might trigger a higher-than-normal rate of cell division and thereby increase the chances of a "genetic accident," leading to cancer.[108]

A second study reported that men with the highest natural levels of circulating IGF-1 had more than four times the risk of prostate cancer compared with those in the reference group. "The association of IGF-1 and prostate cancer risk is stronger than that of any previously reported risk factor, including steroid hormone levels."[109] The same article also cited a 1993 study. Its conclusions regarding the possible links between IGF-1 and osteoporosis in postmenopausal women were "no more encouraging." The study also pointed to "severe side effects" of IGF-1 treatment.

None of this should really come as a surprise to anyone, however. Dale Bauman and M.A. McGuire, at Cornell University, state in a 1995 paper for Monsanto, "We do not fully understand how the IGF system mediates mammary function" and "nutritional regulation of the ST/IGF [BGH/IGF] system appears to be a key component signalling the appropriate use of nutrients."[110]

Given the conversation I had with Brian McBride in 1988 and his statement that they really did not know what the role of IGF-1 was, I still find it difficult to understand how any regulatory agency could assume that the use of rbGH is acceptable, much less safe. The explanation, however, is that they never looked at IGF-1 because Monsanto was not seeking a licence for its use. It was only rbGH that they looked at. This is a choice example of reductionist thinking. Of course you can only find what you are looking for.

The cavalier attitude of the drug industry is expressed by Ray Mowling, Monsanto Canada vice president, in a 1998 letter-to-the-editor, in which he refers to his company's *Posilac* as "the largest selling dairy animal health product in the United States."[111] Mowling did not explain what health benefits this "health product" provides.

Resistance

"We have to demonstrate to the public that we're acting in their interest," warned Monsanto vice president Will Carpenter in 1990. "Otherwise, the handful of vociferous, well-organized critics would dictate what products appeared on supermarket shelves and ultimately what research was allowable in the labs." If the critics prevailed, said Carpenter, "then freedom will suffer, misinformation will reign and the public will be victimized."[112]

The critics were not silent, however. Although some of those who strongly opposed rbGH insisted that its licensing and use were a *fait accompli*, others, myself included, rejected their *de facto* technological determinism and organized against rbGH.

The first item on the agenda of the newly created Toronto Food Policy Council (FPC) in 1990 was rbGH, put there by two of the Toronto regional health committees. It is still an active item on the FPC agenda. In 1991 a coalition of groups and individuals, including dairy farmer Lorraine LaPointe and myself, organized a national Pure Milk Campaign that led to thousands of names on petitions to the government calling for a ban on rbGH and, if that was not achieved, for labelling of the products of rbGH-treated cows.

We made three strategic decisions to guide our organizing and opposition: 1) Stick to the correct name, rbGH, and do not allow the industry to sanitize it to rBST; 2) Do not allow the industry to define the debate as a food safety issue — that's a no-win argument that could absorb all our energy, and at the end of the day we could only prove harm after the fact; and 3) Insist on labelling if the drug is approved. We knew that this last one was a key issue because the biotech industry feared public reaction if dairy products were labelled "rbGH." Interestingly, the dairy processors and their organization, the Canadian Dairy Council, knew this too, and told both dairy farmers and the drug industry so.

There is no doubt in my mind that the sustained, grassroots opposition over the past decade has had a major influence on both politicians and regulators. It also provided both moral support and factual backing for politicians, (some) government regulators, and independent-minded dairy farmers who held out against the pressures of the industry, their own dairy organization, and the technological defeatists. The uproar in the fall of 1998 over the manipulation of research and research scientists by Health Canada, particularly in connection with the review of rbGH, coupled with the determined and highly informative investigation by the Senate agriculture committee into the handling of rbGH, would never have been possible without the public pressure over the past decade.

Finally, several months in advance of Health Canada's stated deadline, a minor ministry official announced on January 14, 1999, that Monsanto's *Nutrilac* rbGH would not be licensed for use in Canada. The news was received with joy around the world, but not by Monsanto.

The decision was based primarily on the report of an expert panel established by the Canadian Veterinary Medical Association as a result of parliamentary committee hearings in 1998. In contrast to this careful and critical report, a parallel report on the human safety aspects of rbGH prepared by an expert panel appointed by the Royal College of Physicians and Surgeons of Canada was essentially a whitewash of the product, repeating all the well-worn generalities of Monsanto's claims and concluding that there were no problems. That panel appears to have ignored all the research I have cited in this chapter.

The veterinarians' report stated:

> The Panel felt that there were a number of legitimate animal welfare concerns associated with the use of rBST. These included an increased risk of clinical mastitis (of approximately twenty-five percent) and lameness (approximately a fifty percent increase in the risk of clinical lameness), and a reduction in the lifespan of treated cows ....

In general, the Panel felt that there were sufficient data available to make a reasonably informed assessment of the effects of rBST. There were four specific conditions (risk of cystic ovaries, twinning, retained placenta, and abortion/fetal loss) for which there appeared to be an effect associated with the use of the drug, but for which there was insufficient evidence to draw firm conclusions ....

For Monsanto the decision was not good news, and Ray Mowling protested that the company had been cheated out of its right to respond to the veterinarians' report before any decision was made by Health Canada. The company then started on one of its familiar press campaigns, pumping out stories for local media. One of Monsanto's news releases was published in *The Prairieland* of Saskatchewan. It quoted Ray Mowling as saying, "The prospect of a final decision being made without an opportunity for a response from Monsanto on the specifics of the science is contrary to normal regulatory approval processes and a basic affront to widely practiced decision making and fairness .... BST has become the largest selling dairy health product in the world ...."

Apparently Monsanto does not recognize the prerogative of Health Canada to make a decision in favor of the health and well-being of Canadian people and animals. The company has yet to describe the health benefits of its performance drug.

Before and after rbGH was approved for use in the United States there were a wide variety of efforts at the state level to ban or to require labelling of rbGH. Small processors that tried to market milk as rbGH-free were met with intimidation and threat of legal action by Monsanto; state legislatures that tried to introduce legislation to ban rbGH or to require labelling of milk from cows treated with rbGH were subject to massive lobbying by Monsanto's hired guns. Essentially Monsanto got its way, though it is possible to find, here and there, dairy products with a contorted label, as required by the FDA, that says something to the effect that the product was produced from cows not treated with rbGH but that there is no difference anyway. In other words, no claims can be made and no negative implications of rbGH-produced dairy products are allowed. It appears, however, that dairy farmers themselves are turning against rbGH as they discover the ill effects it has on their cows. And if anyone (especially Monsanto) knows how many people have reduced their milk consumption or stopped it altogether because they don't want anything to do with rbGH-produced milk, they are not saying.

Europe, on the other hand, has had a ban on importation of meat and dairy products produced by means of artificial hormones of any sort for years. In spite of a 1998 ruling by the World Trade Organization (WTO) that such a ban is illegal according to its rules, the European Union has made it clear that it has no intention of allowing hormone-treated or produced products to be imported.

Chapter Seven

The eternal tomato

"Taste has not yet been evaluated"

I was first introduced to the Flavr Savr tomato by Robert Goodman, vice president of research for the tomato's developer, Calgene Inc., at an industrial biotechnology conference in Toronto in December 1988. Like most presentations on biotechnology, it was intended more to lure investors than to present scientific findings. The most memorable bit of Goodman's presentation was his comments accompanying the slide that purportedly showed the new genetically engineered (GE) tomatoes alongside "conventional" tomatoes after they had all been sitting on a shelf for three weeks.

"What we have seen," he said, "is a dramatic extension of the shelf-life of ripe tomato fruits without loss of other characteristics, at least that have so far been tested. Taste has not yet been evaluated."

This prophetic remark expresses well the priorities of the entire biotechnology project. Taste had not yet been evaluated. Nutrition did not rate a mention. Nevertheless, it is clear that Calgene and its tomato were precedent setters in many areas, preparing the way for the flood of GE crops now being grown and processed into a multitude of foods. For this reason, a close look at its development, and its outcome, is helpful to understanding how we got to the present.

&

When Calgene was established in 1980, said Calgene's president Roger Salquist, there were no plant molecular biologists because nobody had ever engineered a plant.* "The idea was to do the science and get the products developed and registered and market them." At the time, Calgene was developing and marketing conventional cotton, tomato, and canola seed in addition to making frequent trips to the stock market to replenish its working capital.[113]

Calgene described its strategy as building operating businesses "to facilitate the market introduction of genetically engineered products and to maximize the long-term financial return from such proprietary products."[114] Calgene and its backer at the time, Campbell Soup Company, reported on the success of the company's first field trials of the new GE tomatoes in mid-1989. The company chalked up relatively modest losses of $6.8 million on revenues of $30.2. million that year.

Calgene researchers had developed a way to lock up the promoter gene responsible for turning on the production of the enzyme (poly-galacturonase) that produces ethylene in the fruit. (For an introduction to the biology involved, please see the appendix.) Ethylene promotes ripening of the tomato on the vine as well as after picking. Ethylene is also used to "gas" tomatoes picked as "mature greens" before the plant itself has had a chance to turn on its own ethylene production. Calgene did this by isolating the gene, copying it, and inserting the copy "backwards" so that it bound with and effectively silenced the promoter gene, with the result that ethylene production was simply not turned on. Technically, this has been referred to by the industry as "anti-sense" technology. You can decide for yourself what to call it.

In 1990 Calgene received a US patent on its "invention." Then the company asked the FDA for an advisory opinion on the marker gene,** from the antibiotic kanamycin, that accompanied the "antisense" genetic transformation in what it was calling the Flavr Savr tomato. It was the first time the FDA had been asked to evaluate "a component of genetically engineered plants to be consumed directly as whole food."

Less than a year later, Calgene asked the FDA to confirm the status

---

* Unless otherwise indicated, information on Calgene comes from the company's press releases, which used to arrive in the mail almost weekly.

** Since genetic engineering is a random exercise, it is common practice to include an antibiotic-resistant "marker" gene along with the novel genetic material. The resulting organisms can be exposed to the antibiotic of the marker gene. The organisms that survive are those that have been transformed. The killed ones are discarded. See Appendix for details.

of its genetically engineered Flavr Savr whole tomato as food. While working on its tomato, Calgene was also pursuing research in developing canola seed with special oil characteristics, as well as cotton seeds designed both to tolerate a specific herbicide (Rhône-Poulenc's Bromoxynil) that would otherwise kill the cotton plants, and to contain *Bacillus thuringiensis* (Bt), a toxin derived from a naturally occurring microorganism that is lethal to specific larvae, such as the cotton boll weevil. (Calgene owned Stoneville Pedigreed Seeds, a cotton seed producer.)

In mid-1992 Calgene filed its first petition with the US Department of Agriculture (USDA) to grow GE plants commercially without USDA permits. Three months later the USDA ruled that it would no longer regulate the Flavr Savr, meaning that the tomatoes could be grown and shipped anywhere in the US without permits. Apparently the presence of the kanamycin-resistant gene was not considered significant, if considered at all. A few short years later, however, there is growing alarm about the rapidly rising levels of antibiotic resistance as a result of human misuse and the widespread use of antibiotics as growth promoters in livestock and poultry production. The use of antibiotics as marker genes on GE foods is an added and unnecessary burden, not only in regard to antibiotic resistance itself, but also because of their effect as promoters of genetic instability and increased gene flow, the random movement of genes to other organisms.

In spite of its regulatory successes, Calgene's financial losses continued to roll on: $4.2 million was chalked up to development of the Flavr Savr and to buying the marketing rights for fresh market tomatoes from Campbell Soup Co., which, even before the wonder tomato could get off the ground, so to speak, had announced that it did not intend to market any GE product "in the foreseeable future," in spite of its multi-million-dollar investment in the project. "For us to put a new ingredient in our products, not only would it have to be a fully approved product that meets regulatory requirements, but it has to be something we clearly see consumers see as a benefit," said Campbell.[115]

While Calgene was doing its utmost to build public expectations of a great-tasting tomato in mid-winter, other scientists and produce-marketing professionals were taking a different view of the project. The *New York Times* reported, "A growing number of biotech experts contend that Calgene's ability to deliver a superior tomato will have little, if anything, to do with biotechnology. The company's own peer-reviewed research and the work of other scientists indicate that shutting off a single gene to delay softening doesn't have much of an effect on tomato ripening."[116]

At this point, Calgene had already invested $25 million in tomato research and marketing, losing some $83 million in total since its founding, but it was getting the press coverage it so badly needed to keep sucking in new speculative capital. Perhaps it was the "anti-aging" technology that appealed to baby-boomer investors.

The *New Yorker* magazine gave the Flavr Savr mixed reviews in a major article in its July 19, 1993, issue, but Roger Salquist did get the opportunity to tell the world, "We're going to sell a hell of a lot of tomatoes, and the growers, the sellers, our shareholders — everybody is going to get rich."

In the story, author Jeremy Seabrook described his guided tour through a Calgene greenhouse and his guide's explanation of the engineering process. "Our guys just cut the PG gene out, using restriction enzymes. Then we make up an Antisense gene. We attach a kanamycin-resistant gene to the Antisense gene as a marker, and we install this construct in the DNA of a disarmed agribacterium .... We expose tomato cells to the agribacterium, and it injects its DNA, which contains the Antisense gene, into the tomato DNA."

The cells are then cultured into plant tissue and the cultures that have not been successfully transformed are eliminated by subjecting the lot to a dose of kanamycin. The transformed cell cultures, containing the antisense gene and the kanamycin-resistant gene, survive. As Seabrook's guide described it, "We have no control over where the Antisense gene lines up on the genome. Recombinant DNA doesn't give you that control — yet. Sometimes the gene is going to end up in the wrong place on the genome, which means that the tomato will probably develop into an undesirable mutant and we'll have to kill it."

By the time Calgene got this far, however, three of the tomato packers that the company had lined up to grow and ship the Flavr Savr had backed out of the deal, concerned that Calgene's integrated approach of being in control of everything from seed to product distribution would alienate them from the traditional packing business. "Our core business is built on relationships and traditions that Calgene wants to circumvent and compete with. We see them putting our relationship with our customers at risk," said Jim Taylor, one of these packers. "I think they have made some very rosy projections to the world at large and to their stockholders that I don't feel are workable or real."[117]

While waiting for final approval to market the Flavr Savr, Calgene proceeded to market conventional tomatoes under the MacGregor label, but in March 1994 it announced a sharp cutback in this program after losing $3.4 million on sales. It also announced the appointment of former USDA deputy secretary Ann Veneman to its board of directors.

Two months later the FDA gave Calgene formal notice that the Flavr Savr had satisfied its food safety requirements. The FDA's checklist for biotech foods contained four questions:

- Does the food have the same nutrients as other foods?

- Is the food free of toxins?

- Is the food free of proteins that cause allergies?

- Is the food basically like other foods?

If a biotech food meets these criteria, said the FDA, it can be marketed without being specially labelled regardless of the genetic technology used to produce the foods.[118]

The problem with criteria like these, of course, is that they are very vague, and the answers to the questions depend very much on what one is looking for. How much is actually known, for example, about allergy-causing proteins? Or how does one identify novel proteins that turn out to be toxic? The most obvious question, however, is, what does "basically like" mean? (The concept is now in use in Canada as "substantially equivalent." It is discussed further in Chapters 8 and 9.)

To really grasp what was going on, one has to remember that the rules of the game were being made after the game got underway. The pressure to get GE foods onto the market in the hope that a payback on investment would follow was there from the beginning. In 1994 Calgene reported another loss, this time $42.8 million on revenues of $39.4 million for the year ending June 30, 1994. It could not go on losing at this rate forever.

Finally, in May 1994, Calgene introduced its novel tomatoes in selected stores in California and Illinois, but it still had to report a net loss of $30.6 million on revenues of $56.7 million for 1995. Much of this revenue came from the cotton seed sales of its subsidiary, Stoneville Pedigreed Seed Co.

Even though there was little likelihood of the much-vaunted Flavr Savr ever making it north of the border, Health Canada approved the sale of the engineered tomatoes in Canada in February 1995, revealing how eager the department was to welcome "novel foods" into Canada. Health Canada had "compared the Flavr Savr to other commercial varieties and found no difference in composition or nutritional characteristics. Based on Calgene's information, the Department found the Flavr Savr to be as safe and nutritious as other tomato varieties."

There never have been any Flavr Savr tomatoes imported or sold in Canada.

&

Calgene's first business liaison with Monsanto, the company that eventually bought it out, was in mid-1993 when it signed several cross-licensing agreements with Monsanto, thereby resolving a number of patent disputes. Salquist said of the agreements, "This ... will enable both of us to focus on commercializing our products, rather than engaging in costly litigation." This pattern of business cooperation is increasingly popular and takes the form of patent licensing, strategic alliances, partnerships, and joint ventures that help to avoid competition while also increasing business concentration.

In spite of all the investment and promotional efforts, Mother Nature remained unimpressed with Calgene's wonder tomatoes, and Calgene had to report early in 1996 that "most of the Flavr Savr tomato varieties that Calgene had available for production did not have acceptable yield and disease resistance performance. Consequently, Calgene plans to temporarily limit its tomato growing operations beginning in the spring of 1996 until it is able to complete its development of Flavr Savr varieties that have enhanced commercial agronomic qualities."

"Monsanto Co. agrees to throw Calgene Inc. a financial lifeline" was the *Wall Street Journal's* headline in mid-1995 when Monsanto paid $30 million in cash and its interest in NT Gargiulo, the largest packer and shipper of fresh tomatoes in the US, for a 49.9 percent share of Calgene [119] Monsanto also agreed to provide "long-term credit facilities for the general business needs of Calgene and Gargiulo."[120] In other words, Monsanto bailed out a sinking Calgene.

After its shareholders had approved the deal, Calgene announced, "The marriage of proprietary Gargiulo tomato germplasm with Calgene and Monsanto technology will give us a substantial advantage in becoming the low-cost fresh tomato producer and the first true nationally branded tomato business."

A few months later Calgene announced that it would quit trying to grow its Flavr Savr tomatoes in Florida and, instead, concentrate on conventional tomatoes with NT Gargiulo. Calgene had found it too costly to harvest the Flavr Savr tomatoes as they ripened. The tomatoes were too soft for mechanical harvesting and the labour required for hand picking was too expensive, so Calgene planned to grow its GE tomatoes in Mexico where labour was cheaper.

At the end of July 1996, Monsanto made a further $50 million equity investment in Calgene, giving it 54.6 percent ownership, and Roger Salquist resigned from the positions of chairman and CEO. Six months later Monsanto acquired the remainder of Calgene for $240 million. After that the tomatoes slipped out of sight over the horizon into the sunset. What Monsanto got for its money was a lot of basic science in plant breeding and genetics, including herbicide resistance and Bt "technology," plus their applications in canola and cotton.

&

In retrospect, Calgene's 1994 hype for its mythical MacGregor tomato is just a bit pathetic:

> The press coverage of the introduction of the MacGregor's tomatoes grown from Flavr Savr seeds was extensive, nationwide and overwhelmingly positive. We estimate we had more than 102 million gross media impressions in the first 72 hours after clearance. This remarkable coverage of a new food product has established the MacGregor's brand firmly and positively in the minds of consumers, and even more importantly, in the retail grocery industry. Almost all the press coverage focussed on whether our tomatoes taste significantly better than the standard supermarket tomato. The near unanimous

response is that they do! The threats and bombast of the biotech opponents have proved to be hollow and now seem largely irrelevant.[121]

Looking back, I realize how modest Calgene's message had always been. Roger Salquist's claims for his tomatoes may have been bombastic, but he offered no flamboyant prophecies of impending apocalypse if genetic engineering was not allowed to transform our food supply tomorrow, and no wild claims about how genetic engineering will save the environment.

GE foods have come — or been pushed — a long way since then.

Chapter Eight

Killer potatoes

While Calgene said it was trying to make a better tomato for eating, no one was making a similar claim for the transgenic potatoes under development by Monsanto. The best the company could do was make specious claims about how the Bt potato would be better for the environment because farmers would be able to use fewer insecticidal sprays.

The problem for which transgenic Bt potatoes are supposedly the solution is the Colorado potato beetle, a major pest for potato growers. The real problem, however, is the massive monoculture production of potatoes in certain limited geographical areas such as Maine and New Brunswick, Manitoba, Idaho, and Washington, which sets out a massive banquet for predators. But there is more money to be made selling high-priced, GE seed potatoes to big industrial farmers than there is in working toward a more highly diversified, and thus less concentrated, agriculture in harmony with its environment.

Transgenic Bt potatoes, in other words, are the wrong answer to the wrong problem. They are, however, the right answer to the problem of corporate profits.

Even in diversified ecological systems there will always be some pests, but there are other approaches to dealing with them. For example, in 1992, researchers at Cornell University were playing with a hybrid potato bred from a wild type, *solanum Berthaulthi*, that has thin hairs on its foliage. This potato has a striking ability to ward off pests thanks to these hairs, which secrete a sticky substance that traps and kills small insects such as the leaf hopper as they feed or reproduce. These plants also know how to deal with the Colorado potato beetle. The beetle gets a serious case of constipation from the sticky secretion, which causes its stomach to bloat, crushing its ovaries and curtailing its reproduction. Robert Plaisted, the Cornell professor doing the research, said the potato tastes like any other but comes equipped with the best method yet of providing a broad spectrum of resistance to insects. Plaisted says this "new" potato has found favor mainly with organic growers. He hopes to have other varieties with similar characteristics available in the new millennium.

When I asked him about transgenic Bt potatoes, Plaisted told me that one of the problems with them is that they offer no deterrent to the leaf hopper, which is actually a bigger problem than the potato beetle because the hopper is very small and the damage is done before the farmer realizes there is a problem. When I asked whether the fascination with genetic engineering was affecting his research, he replied that fortunately he received special grants from the USDA and from an international foundation. Without those his work would not be possible.[122]

In Ontario, potato growers have good success with a very simple bit of technology: the potato fields are ringed with shallow trenches lined with plastic. When they hear the dinner bell, the beetles head for the potatoes from their refuges in undisturbed areas, slide into the trenches, and cannot crawl out again. (The dead beetles might be a good protein supplement for chickens.)

This provides a good illustration, and there are many, many more, of the potential alternatives to both conventional agrotoxins and genetic engineering. But Plaisted's hairy potato won't require specific pesticides and will not sell for a premium price plus a "technology fee" for the privilege of growing the potatoes the way Monsanto's GE products do. And the plastic can be used year after year.

&

Canola, corn, soybeans, potatoes, and cotton, all major crops of industrial agriculture, were chosen for genetic engineering for three major reasons: 1) The volume of production, and hence the volume of GE seed required, could pay for its development and provide an attractive profit to the corporations; 2) Their structures lent themselves to genetic engineering, while other crops, such as wheat and rice, have been much more difficult to manipulate satisfactorily, or profitably; and 3) They are all primarily raw material for manufactured or processed end-products.

The potatoes can and will go to companies like Nestlé for processing into french fries for McDonald's or into potato chips.* The canola seed will be processed into an edible oil and animal feed. Soybeans are likewise processed into oil and meal, to reappear as an ingredient in one form or another in a multitude of products. The corn is used as animal feed or is processed into corn oil, high fructose corn syrup, starch, and so on. Cotton produces fibre, of course, but also cottonseed oil and meal, and again the meal goes into animal feed while the oil may be used for human consumption.

The public is not, in other words, confronted with a large number of identifiable GE whole foods. (Potatoes are the exception.) Instead, a high percentage of the processed food products on the market may contain, or be made from, these engineered foods. Take a good look at the labels of a random selection of processed food products: corn and soy, in one form or another, may be found in more than half the products. Lecithin, for example, is a soy product.

It is not an accident that there is no labelling to advise the customer that the product is produced through biotechnology. If we ask, we may be told that the canola oil is not transgenic, just produced from transgenic seed. Or we are told that it is impossible to keep the transgenic seed segregated from the rest. There seems to be a whole cupboard full of excuses as to why foods produced through genetic engineering cannot be properly identified as such.

&

Potatoes are grown practically everywhere by home gardeners. They are also a major processing crop. In Canada and the US the russet Burbank** is the favorite processing potato. It is therefore, not surprising that the russet Burbank appeared on the market as the first trans-

---

* Midwest Food Products Inc. in Carberry, Manitoba, for example, is actually owned by Nestlé and produces all of McDonald's french fries in Canada.

** The B is capitalized because the variety was developed by Luther Burbank.

genic potato, engineered to contain *Bacillus thuringiensis* (Bt) toxins derived from a common soil bacterium.* That it also contains a gene conferring antibiotic resistance, as a marker of the successful genetic transformation used to develop the Bt potatoes, cannot be overlooked, particularly when there is no longer any technical reason for its being there.

The Bt toxin, as mentioned before, is lethal to the Colorado potato beetle. "Bt toxins kill by binding to a receptor molecule in the insect's digestive system and causing the gut wall to disintegrate; insects can become resistant by evolving a differently-shaped receptor that no longer binds the toxin."[123]

As it turns out, it is not just any Bt gene that will do the trick. Part of the "magic" of Bt is that each strain contains a toxin that is highly specific to certain lepidopterean pests, while leaving every other organism unharmed — apparently. This is the reason that natural Bt extracts have been used by both organic and industrial farmers for the past thirty years or so in the form of foliar sprays (applied to the leaves) applied to everything from potatoes to spruce trees (to fight the spruce budworm). Applying it externally when required, even by air over forests, is, however, a vastly different proposition than engineering it into the potato, corn, or cotton plant so that it is present at all times throughout the plant.

The characteristics of Bt enable it to be praised, and sold, as an environmentally superior way to deal with potato beetles because, we

---

* "Unmanipulated Bt bacteria produce a crystalline protoxin. This protoxin is a preliminary stage of the toxin. It is only in the alkaline mid-gut of the insect's intestines that this protoxin is decomposed in several steps (up to seven) and turns into the actual toxin (which is a much shorter protein that the protoxin). The crystalline protoxin has only specific effects on particular insects. It is very sensitive to UV radiation and decomposes quickly if not eaten.

"By contrast, genetically manipulated maize plants, which contain an artificial, truncated Bt-gene, produce a toxinlike protein that is already about half the size of the bacterial protoxin. It only needs a minor step to turn it into the active toxin. Also, it seems that there is no need for an alkaline value (a high pH) for this step to happen. And finally, the truncated Bt-gene in maize is responsible for three more Bt proteins. Whether any of these Bt proteins acts as an already active Bt toxin is not known and no further examinations have been made.

"It is therefore highly probable that the toxinlike protein of the manipulated maize can be activated if an insect's intestines do not have a high alkaline value. As a result, the toxin might also have effects on earthworms and other insects [and humans], the intestines of which do not have a high pH." (Florianne Kochlin, "Genetically Modified Bt maize An Ecological Risk," distributed electronically, 17/2/97) are told, it is target-specific, non-polluting, and degrades readily.

To say the "seed" is sold, however, is misleading, because in a sense the owners of the technology, the seed companies, do not sell it at all; they rent it out to the farmer for a season. The farmer is not allowed to keep any of the crop for replanting or to share it with a neighbor because the technology, and hence the potato, are owned and patented by one or another major transnational corporation. In the case of the Bt potato, the corporations are Monsanto, Novartis, Mycogen (now a subsidiary of Dow Chemical Co), or some combination of them. So the Bt technology is quite like an externally applied insecticide in that the farmer pays for its use and uses it up. There is no residual value to the farmer/grower. Any residual value flows back to the owner of the technology in the form of profits and control.

There are other major issues with Bt. Most significant is the certainty that its utility will be short lived: such universal and continuous exposure to the toxin will apply tremendous selection and adaptation pressure on the pests; the population of those that are naturally resistant will rapidly expand and others will find ways to adapt their organisms to their new climate, making the Bt toxin useless, not only to the growers of industrial potatoes, but to everyone else who has been and could be using it as a topical insecticide in its natural form.

In addition, the claim that the toxin degrades readily is based on two false premises: the isolated toxin does not degrade as readily as its complex natural form, and the degradation depends entirely on the health of the soil and the population of microorganisms in the soil that are there to break it down. In soils that have been virtually sterilized with pesticides and synthetic fertilizers, the microorganisms to break down the potato plant residue and the Bt toxin it contains are not present.

⅋

I have a vivid memory of hearing a Monsanto salesman make his pitch for the wonders of Bt-enhanced potatoes to a large biotech conference several years ago while the guy next to me, from one of the major agro-toxin companies (maybe DowElanco) was muttering, "That's got to be the stupidest thing I've ever heard. Everyone knows that if Bt is everywhere it will only be two or three years before every bug is resistant to it."

The phenomenon of genetic adaptation, or the acquisition of the ability to withstand normally fatal toxins, is not new nor unknown.

Certainly at least some of the scientists working for Monsanto have known all along that if they put Bt into every seed potato they can, it will be only a matter of time before the target insects learn to cope with this new element in their environment. Since there is no such thing as genetic uniformity in a wild population, there will be survivors, and their population will expand so they become the "normal" potato beetles, rather than their dead cousins who did not carry that ability and did not adapt in time. Or, it is now being realized, organisms can create the resistance from scratch, as it were, quite possibly aided in this by the antibiotic marker genes now also present in their environment.

It was already known in 1993 that a handful of pest populations had grown resistant to Bt toxins — something many pest-control experts thought would never happen — and that more than 500 species of insects showed resistance to at least one chemical. Bruce Tabashnik, an expert on resistance at the University of Hawaii, first found Bt resistance in 1985. "Despite more than 30 years of fighting insect resistance to one insecticide after another, scientists know little about which tactics work and which don't. Most of what they do know comes from theoretical calculations, computer models and a few artificial laboratory experiments." One aspect of the problem, according to Tabashnik, is that "whenever there's a new insecticide, people think of reasons why it's impossible for insects to become resistant to it." This is illustrated by Monsanto, Dow/Mycogen, Pioneer, and Novartis, which have proceeded to engineer potatoes, corn, and cotton to contain Bt, but have failed to develop any coherent and workable strategy even to slow down the target insects' acquisition of resistance. The only generally agreed-upon strategy to limiting resistance is to grow toxin-free plants in refuge areas, or refugia, alongside the Bt variety. "Refuge areas would allow pest individuals that are still susceptible to Bt toxins to survive and contribute their genes to the next generation."[124]

In May 1995 the US Environmental Protection Agency (EPA) approved the *Bacillus thuringiensis* delta endotoxin CryIII(a) for use in potatoes. Three months later the EPA gave conditional approval for the "full commercial use" of Bt CryIA(B) delta endotoxin to combat insect damage in field corn. In announcing the approval of Bt in corn, the EPA said that it had reviewed and approved the resistance management plans for Bt corn submitted by Ciba Seeds and Mycogen and concluded that they "would reduce the possibility of resistance developing for three to five years following use of the corn plant-pesticide."[125] There was no mention of potatoes.

In October 1995 Monsanto received final EPA approval for commercialization of Bt cotton, or what Monsanto referred to as "insect protected cotton containing the Bollgard™ gene." Canada tagged along, approving Monsanto's Bt-engineered NatureMark potatoes in January 1996 and its Bt cotton a few months later. The "Decision Document"[126] for the potatoes notes that the plants were transformed with genes conferring resistance to the Colorado potato beetle (CPB) and resistance to kanamycin as a selectable marker, and reported that it had "determined that these plants with novel traits should not pose concern to environmental safety." Health Canada had already determined, in August 1995, that "food derived from these potatoes is substantially equivalent to that derived from currently commercialized potatoes."

In spite of describing the kanamycin gene used and its function, there is no mention of its possible presence in, or transference to, other organisms, including humans. The document states that the gene "degrades rapidly in simulated mammalian gastric and intestinal fluids," but this does not necessarily mean that we won't absorb the kanamycin-resistance gene directly through our intestinal wall, or that there are not other forms of transference or relocation of the alien DNA directly or indirectly. (One has also to ask, were these "simulated gastric and intestinal fluids" comparable to those of a middle-aged white male or a pregnant black woman or a young Asian child?)

The Decision Document foresees the development of target-pest resistance because "Target insects will ... be exposed to significantly higher levels of Bt than through the current foliar spray treatments, leading to high selection pressure for resistant CPB individuals." Nevertheless, the agency responsible for regulating the processes and products of biotechnology could only say that it "understands NatureMark Potatoes has developed and will implement a pest resistance management plan." In elaborating on this, the document merely states that there are to be "education tools for growers" and the "promotion" of integrated pest management practices. As well, "Detection of confirmed resistant CPB populations will immediately be reported to

AAFC [Agriculture Canada] and a procedure for control of resistant individuals must be available for immediate action." It is hard to comprehend how, on the basis of "sound science," in the absence of any definite and compulsory resistance management plan, and with the limitations of knowledge and experience expressed even in this document itself, approval could be granted for the commercial, unconfined production of Bt potatoes. *Why* approval was granted, on the other hand, is obvious: the job of the regulators is to approve new products for market (see Chapter Nine).

In 1998 AAFC gave approval to the unconfined release of two more lines of Bt potatoes engineered by Monsanto: Atlantic and Superior. The language of Decision Document 67-20 is notable: "Monsanto has provided data ...," "data was presented ...," "the data provided showed ...," "AAFC evaluated data submitted by NatureMark .... " It is impossible not to notice that all data for evaluation was supplied by Monsanto and typically only "evaluated" by Agriculture and Agri-Food Canada / Canadian Food Inspection Agency. (We have already discussed the pitfalls of this approach in the case of rbGH).

AAFC's determination from its survey of data provided by the supplicant was that "the unconfined release of these NewLeaf™ Atlantic lines, when compared with currently commercialized potato varieties, will not result in altered impacts on interacting organisms, including humans, with the exception of CPB and potato flea beetles."

The sloppiness, or indifference, of the regulatory process is reflected in a reference to "NewLeaf Atlantic potatoes previously authorized for release" when it was russett Burbanks that had been authorized the year before. One's confidence in the regulators is not restored by statements such as "AAFC ... concludes that gene flow from NewLeaf Atlantic lines to potato relatives is not possible in Canada," when the instability of DNA and the frequency of gene flow is known to be essential both to ecological stability and the survival of the species itself. As Steven Rose writes, "The essence of the stability of the whole is that the individual components are in constant flux. Freeze them in reductionist immobility, and ... the cellular edifice would collapse into those individual components that we biochemists have for so long lovingly studied in dissected and impoverished isolation."[127]

In 1998 AAFC still "understood" that NatureMark Potatoes "has developed and will implement a pest resistance management plan." The same sweet hopefulness characterizes the regulation of Bt corn, which Monsanto and others had been working on. Resistance management has been a much hotter issue for Bt corn than for Bt potatoes, probably due to the much larger amount of acreage planted to corn.

In 1995 Mycogen Plant Sciences responded to criticism from the Union of Concerned Scientists, saying the "cornerstone" of its resistance-management strategy was to "maintain substantial refugia for susceptible insects with a combination of market forces and grower practices." What it meant by "market forces" was that since Mycogen and Ciba Seeds together produce only about 5 percent of the seed corn in the US, and since an individual farmer will not typically plant all acreage to one particular hybrid, there will be more than enough non-Bt corn grown to provide adequate refugia. Mycogen then slyly commented that "It is incorrect to think that EPA would approve the deployment of these crops prior to the demonstration by seed companies that effective resistance management strategies are in place."[128]

Nine months later a Ciba Seeds spokesman, Dr. Bruce Hunter, told the Ontario Corn Producers that insect resistance was a possibility, though not a known problem. Because the corn borer (a larva, like the CPB, at the crucial stage) is fairly mobile, he said, it continues to breed with the wild population and that would keep resistance from developing.[129]

Another nine months passed (the period in question now spanning two crop seasons) and another Mycogen spokesman admitted that a major question on everybody's mind was how much non-Bt corn had to be grown to ensure bugs continued to be susceptible. "Companies have invested a whole lot of money. If resistance does develop there'll be some very angry shareholders." Bert Innes was, nevertheless, pinning his hopes on the prediction that the premium price of Bt corn ensured that some growers would always opt for the cheaper conventional varieties.[130] This is presumably what Mycogen meant by "market forces" — a rather casual and unusual resistance management plan, to say the least.*

---

* In October 1998 Bert Innes was appointed by Agriculture Canada to the new position of strategic analyst, Biotechnology.

Later in 1996, Northrup King/Sandoz Seeds was advertising its "YieldGard insect-protected corn by Monsanto," telling farmers that when they plant Bt corn, "You don't change the way you do things, such as adjusting your planting schedules, crop management or herbicide program."[131] Soon after, I noticed the same ad on the Sandoz Seeds booth at a farm show. I asked the agronomists attending the booth if there was not a requirement for a resistance management plan. To my surprise they smiled and said yes, and agreed that development of resistance to the Bt toxins was a real problem. But, I said, your ads say explicitly that nothing like that is needed. Their reply was distressingly familiar: There is not enough Bt corn on the market yet for that to be a problem, and by the time there is enough Bt corn being planted for the problem to arise, we figure that management plans will be in place.

By March 1997 Sandoz and Ciba Geigy had mutated into the single organism Novartis, and a company spokesman repeated the line that the companies did not expect a problem for at least a couple of years since only about 10 percent of corn acreage in Ontario would be planted to Bt hybrids. Mycogen and Novartis were still saying that since only 5 to 10 percent of the corn would be Bt engineered, there was not a problem, but Pioneer Hi-Bred and DeKalb were both requiring growers to sign management agreements when they purchased Bt corn. The agreements bound them to plant a minimum of 5 percent of their corn acreage with a non-Bt variety as refuge and not to spray it with corn borer insecticides.[132] Whether 5 percent is enough is another question.

In mid-1998 Pioneer stated, "We require our customer to sign a Gene Technology Agreement that outlines the terms of our insect resistance management strategy and indicates that they understand their responsibilities under that agreement. If the grower does not routinely treat with an insecticide (and few in Ontario or Quebec do), we recommend that they plant up to 95% of their corn acres with Bt corn and a minimum of 5% of their corn acres with non-Bt corn to serve as a refuge...Theoretically at least, the purpose of that refuge is to maintain a small population of corn borers nearby that are susceptible to the Bt protein found in our products and available to mate with any resistant mutants. The non-Bt hybrid chosen should be agronomically similar to the Bt hybrid, planted at the same time and in an adjacent field (the neighbor's field does not qualify) to the Bt hybrid."[133]

Novartis never did have a management plan. In a September 1998 press release Novartis crowed that three years after introducing the crop they were finally insisting that farmers follow a management plan — by promising to pay them to do so in the future. Under their "Bt Stewardship Program," according to the press release, "the financial incentive varies based on the quantity of NK Brand YieldGard or KnockOut corn seed purchased. Growers who buy a significant amount of Bt seed will receive substantial savings if at least 20 percent of their order includes non-Bt hybrids. With this program, we're offering to share IRM [insect resistance management] stewardship responsibilities with our customers so we can preserve this technology for years to come." Note that the refuge area has grown from 5 percent to 20 percent!

In response to Novartis's notice, the US National Corn Growers Association (NCGA) said it feared that the EPA could restrict insect protected crops to the point that they would no longer be economically practical for farmers. "We want to utilize these hybrids for the economic and environmental benefits they bring to corn production," said NCGA president Ryland Utlaut, who worried that the EPA might establish insect resistance management programs that corn growers would find too expensive to implement.[134]

At the same time, an Ontario corn grower was telling the annual meeting of the Canadian Crop Protection Institute not to waste its time trying to sell the idea of refugia to farmers. Farmers were not going to pay a premium price for Bt corn hybrids and then plant "junk" hybrids to create refugia. Farmers would depend on neighbors "who are not so progressive" to plant older hybrids and provide the refugia, he said.[135]

&

In April 1996 the USDA sponsored a National Forum on Insect Resistance to *Bacillus thuringiensis*. The "ideas and actions suggested by participants" were published in the February 1998 issue of *Nature Biotechnology* after a thorough review by all the participants. The author/participants pointed out that what knowledge there is about resistance management "comes mostly from laboratory and small-plot research," and that "large-scale experimentation is probably needed to resolve some of the questions." *"Limited understanding of the behavior and biology of pests is one of the most serious knowledge gaps.* A much more detailed understanding of pest movement between plants, fields, other crops or weeds, and geographical areas is needed to effectively deploy refuges." (emphasis in original)

The report also identified the twin problems of who should pay for the necessary research and who should be responsible for the "effective monitoring" required of both research and practical application. "Clearly, the appearance of a basic conflict of interest would be present if certain aspects of the monitoring were controlled by industry."

Despite these important concerns, in the two-and-a-half years that elapsed between the forum and its report, Bt corn, potatoes, and cotton have been planted on an ever-increasing scale with no particular resistance management plans or practices in place.

As *New Scientist* put it, "The panel warned that unless the EPA forced companies to take action, the pesticide could quickly become worthless .... The panel failed to make specific recommendations, arguing that no one knows enough about resistance to prescribe specific measures."[136]

Fred Kirschenmann, a respected organic farmer in North Dakota, commented, "If insects are a permanent part of the landscape...then shouldn't we begin thinking about insects as permanent residents rather than disposable pests?"

> If we know little about the complex ecologies of which insects are a part, doesn't the introduction of this technology pose the possibility of additional mischief that we have not yet thought of? Admittedly, entomologists know a lot about the behavior of individual insect species in the controlled environment of the laboratory and the research plot. But we still know precious little about the deeper ecology of complex interrelationships of species within local ecosystems especially with the added uncertainty attendant to the introduction of novel organisms into ecologies with which they did not evolve ....What else, besides resistance, may we be unleashing into the environment? Why do we want to incur this risk when there is such a questionable, short term benefit at stake?[137]

The most recent assault on the potato is the effort to transform it into a vaccine carrier. Scientists at Boyce Thompson Institute at Cornell University have inserted a gene for a bacterial antigen — the protein that stimulates the production of protective antibodies — into potatoes.

They are hoping to "wipe out" enteric diseases, such as diarrhea and cholera, by developing vaccines grown in specific foods, such as potatoes or bananas, that need no refrigeration. In their first study, the researchers used a gene from *E. coli* that can act as a vaccine against travellers' diarrhea, which comes from intestinal infection caused by contaminated food or water. While tourists commonly contract the bacterium in the Third World, it is a much more serious health problem for inhabitants of those countries, and diarrhea caused by bacteria is a major cause of infant mortality worldwide.[138]

As in every other instance, one does have to wonder if addressing the underlying problem — in this case contaminated water — might not be more to the point. Now, as well, the wisdom of attempting to "wipe out" pathogens is itself being questioned. "The millions of viruses, bacteria and fungi that kill or blight plants worldwide should be conserved with the same urgency as other species, says a group of biologists*. Alarmed that many are being eradicated, the scientists launched a campaign to 'preserve the pathogen'" at the Seventh International Congress on Plant Pathology in Edinburgh in August 1998. With the rapid loss of habitats and ecosystems worldwide, the increased use of fungicides, pesticides, and herbicides in agriculture and the release of genetically modified organisms, the threats to pathogen diversity in the wild are immense, said the biologists.[139]

&

A very different, and perhaps even more alarming, problem is the effect on humans who ingest Bt, whether indirectly through products manufactured from Bt corn, or directly from eating Bt potatoes. Theoretically there is no problem because Bt is inactivated in acidic environments, such as the normal human gut — but not everyone is "normal." This has never been addressed because reductionist science, the corporate drive to get new products to market, and the regulatory agencies' sim-

---

*	Total estimated number of species	Known number of species
Bacteria........3 million 4000 (0.1%)		
Fungi..........1.5 million 70,000 (5%)		
Viruses.........0.5 million 5000 (5%)		*(New Scientist, 22/8/98)*

plistic conceptions of life see no reason to address it. Their negligence is obscured by the faulty concept of "substantial equivalence." In common parlance, this concept means that despite the introduction through genetic engineering of traits, such as a Bt toxin, that are not found in conventional potatoes, the resulting potato is, for all intents and purposes, including human consumption, substantially equivalent to a normal potato. Of course "substantially" is a rather vague and subjective term, but it is precisely that vagueness that is useful if a regulatory agency is predisposed to overlook the novelty of genetic engineering.

Obviously there is no such thing as a standard-issue or "normal" stomach. Nor is the toxin inserted into the potato the whole structure of Bt as found in the soil or in normal use in non-engineered foliar Bt sprays, which are composed of the complete Bt bacterium. What is inserted is the isolated end-product toxin produced by the Bt bacterium. This "naked" toxin is expressed throughout the plant, including in the tubers. But having decreed that Bt potatoes are *a priori* "substantially equivalent" to conventional potatoes, the regulatory agencies have said there is no need for clinical trials to test the actual effects of Bt potatoes on a range of human beings.

There is, however, anecdotal evidence to suggest that human stomachs that are not highly acidic can have some of the same reactions to Bt as the Colorado potato beetle. Complaints to Monsanto from persons apparently affected by Monsanto's potatoes have met with denial and the threat of a liability lawsuit since the victim cannot prove anything. It is, after all, hard to prove that a potato has poisoned you when you have eaten the evidence, which was not labelled in the first place.

While the seed was being sold and planted, and the acknowledged issue of resistance management was given little more than lip service, at the retail end of the food chain the Maritimes' largest retail grocery chain was engaged in a promotional campaign with Monsanto for NatureMark Bt potatoes. This was in the spring of 1996, which means that the potatoes had been grown in 1995.

Monsanto supplied an attractive customer handout carrying the quaint NatureMark logo, with its outline, in color, of a person hoeing a crop and the words "NatureMark potatoes — grown the better way." Inside, the "better way" is described: "These plants have a built-in protein taken from nature which is part of a family of proteins used by home gardeners for more than 30 years. So these plants need only sunlight, water and fresh air to protect themselves from the devastating Colorado potato beetle." (The designer forgot about soil.) The potatoes were packaged in attractively printed plastic bags displaying, in full colors, similar language.

For its part, Sobey's, the chain selling the potatoes, put out a press release that said, "We're proud to offer our customers a choice of premium NatureMark Russet Burbank potatoes that are better for the farmer and the environment." The press release also repeated the Monsanto script that "due to their natural protection from the Colorado potato beetle, NatureMark potatoes use less energy and create less waste, which is better for the environment," but there was no explanation as to how this was possible. NatureMark was identified as "a potato seed company based in Boise, Idaho." Monsanto was not mentioned.

It turned out the potatoes had been grown without Canadian environmental approval. "A 'loophole' in federal regulations allowed experimental crops of genetically altered potatoes to enter the marketplace last year (1995) without the knowledge of environmental regulators ...NewLeaf potatoes were mixed into regular potato supplies beginning last November before they had received environmental approval and before consumers had been informed ....[They were] approved for human consumption by the federal Health Department in November .... NewLeaf received environmental approval in January [1996]. According to Margaret Kenny of Agriculture Canada, trial crops do not require environmental approval since they are grown under controlled conditions."[140] Apparently these "controlled conditions" allow for sufficient production for trial marketing on a substantial scale.

The newspaper article reporting this elicited an immediate response in the form of a letter to the editor from Frank Claydon, senior assistant deputy minister of the department of agriculture, in which he said that the article "needlessly alarms consumers as to how well the government can regulate the food they eat." He called the suggestion that there is a loophole "unfounded." "Canada has one of the safest food systems in the world. An integral part of this system is that new biotechnology products must meet stringent requirements for human, animal and environmental health and safety."[141] Unfortunately, Claydon did not detail the "stringent requirements." And as I have shown, public alarm is quite appropriate.

Chapter Nine

Made to order: Regulation

The time required for regulation ... must be weighed against the time dimension of capital. — Richard Mahoney, president, Monsanto[142]

There is a fundamental contradiction between the culture of technological determinism and the meaningful regulation of the processes and products of technology in the public interest. In this context, regulation winds up as a PR function to provide assurance to the public that regardless of its perceptions or experiences, all is well and there is no cause for concern or alarm. In fact, since in this culture it is believed that technology serves the public interest by definition, any public criticism is just inappropriate.

The now-defunct Science Council of Canada pointed out back in 1982 that "the powerful economic incentive for industrialized nations to seek and exploit technological innovation, even in the face of recognized scientific uncertainties and ethical dilemmas," constituted a substantial obstacle to meaningful regulatory strategies. In light of this, the Council recommended that "the public should have more say in social and policy decisions, and in turn those who make regulatory decisions should be more accountable to the public."[143] The government response, eventually, was to treat the Science Council as a noxious weed and eradicate it.

Process or product?

Perhaps because of Canada's unequivocal support of novel technologies, particularly nuclear and genetic engineering, the biotech industry has chosen Canada to be the lead player in the race to bring new and untested products to market. This choice could also have been influenced by the total absence in Canada of any tradition of public policy debate concerning either agriculture or technology. Not only is there no tradition of debate, there are no mechanisms or fora in which debate and policy formation might take place.* What organized "discussion" there is comes from the private industry-financed organizations such as the Fraser Institute that shape market policy and are used by the media to explain business-centred public policies to the public.

In this context, the market-obsessed government of Canada has created a situation in which the ministries of health and agriculture are expected to pursue the contradictory policies of promotion and regulation. This contradiction is particularly obvious regarding technology in general and biotechnology in particular. A third ministry, industry, does not face this problem since it is expected uncritically to promote any and all technology, particularly the death sciences of nuclear energy, drugs, and biotechnology, as "drivers" of the national economy.

The contradictions between promotion and regulation have become increasingly acute in recent years as the government has endeavored to duck out of its public responsibilities in favor of serving its corporate "clients" and meeting their expectations that the government will facilitate their plans to get new products to market and dividends to shareholders.

This peculiar approach to industry regulation developed with the rise in influence first of the drug industry and then the drug and biotechnology industries combined, which are now for the most part the same foreign transnational corporations. Pharmaceuticals, medical research and agricultural research, and production now form a continuum.

Historically, the regulation of foodstuffs has focussed on contamination and adulteration (by way of additives and extenders) — both

---

* The Science Council noted this in its 1982 report. "In Canada, no forum exists to discuss concerns about this new technology along with the promises it holds."

physico-chemical properties — and Canada's food regulatory regime reflected these priorities. In other words, the emphasis has been on how the food is treated — the process through which the food has passed — on the way from farm to grocery store. The fundamental issue is whether the food has been treated with respect for what it is or with contempt as simply a commodity and a means to make money. Adulteration and contamination, and now genetic engineering, are, not surprisingly, consequences of the latter attitude.

The focus of the regulation of foods has been "safety," on the assumption that food is healthy and "safe" to start with, but needs to be handled with care and processed in appropriate ways, without contamination or adulteration, the farther it has to travel. The introduction of pasteurization of milk and a concern with sanitary conditions in processing plants, for example, were focussed on the purity, and thus healthfulness, of the product as the urban centres grew and the distance between cow and consumer grew accordingly. While the effectiveness of regulatory requirements could be measured by the wholesomeness of the product, the regulations and the achievement of legal purity focussed on the process by which the food was produced.

Pasteurization is a process of sterilization, required by law, with strict specifications. (Whether they are the best measures to be pursued to ensure that milk is healthy, or "clean" as some would say, is a matter for discussion elsewhere.) Pasteurization is not a product, though its effectiveness can be measured by the quality of the product. The fact that it is the processing that is considered crucial is reflected in the law that prohibits the sale of unpasteurized milk regardless of its quality or wholesomeness.

In the case of meats, regulation has also focussed on process, and standards for processing and handling, as well as building standards and allowable materials, have been stringently laid down. The assumption has again been that if the process is sound and sanitary, the product will be healthy or pure. This has been monitored by trained inspectors who examine the product superficially (smelling and visual inspection) after slaughter. These inspectors have always been government employees since their function is to protect the public.

This approach has been deemed increasingly inadequate in recent years because it has not prevented instances of serious and widespread food poisoning. In meat, in particular, there have been serious short-comings with an inspection system that assumes the process by which the animals have been raised is sound, and that any health or safety problems are related to specific animal defects and diseases. (Health problems are personal, not environmental, in other words.) The prolif-eration of pathogens, however, is precisely a problem of production methods, from intensive, confined hog and poultry operations and beef feedlots to the massive high-speed processing plants, each of which are pathogen-production (or replication) areas. In fact, it is the density and speed of operations — from poultry or pig barn and feed-lot to kill line and processing — that are major causes of pathogen pro-liferation.

The response of the US and Canada has been to adopt new HACCP (Hazard Analysis Critical Control Point) food inspection procedures. Under this regime, the end product will be examined even less than previously, the theory being, as before, that if the process is sound, the product will be as well. The procedure establishes a number of "critical control points" where contamination (introduction and/or spread of pathogens) is most likely to occur, and these are monitored. As long as the procedures at these points are considered sound, the product is presumed to be safe. It is to be applied from farm to retail (eventually, in theory), but there is no indication that the character of the processes themselves will be questioned. Feedlot density and feeding regimes; confined poultry and hog operations; and intensive crop monocultures, including the use of agrotoxins, from pesticides to antibiotics and hor-mones, will continue to be regarded as the only efficient way to produce industrial commodities, and therefore beyond questioning.

There has been a great deal of debate about the effectiveness of these new procedures and their cost, particularly for smaller plants — "The purpose of the final rule [on HACCP] is to improve food safety, not cause the meat and poultry industry to consolidate," commented the editor of *Meat & Poultry*[144] — but the issue remains obfuscated by the refusal of anyone in the industry to look upstream to the possibility that the problems originate in primary production. Attention remains focussed downstream on how the meat is handled as it speeds by the meat cutters on the dis-assembly line.

The regulation of irradiation, which the nuclear industry has been pushing since the end of the Cold War as a means of "sanitizing" food, ("cold pasteurization" they are now trying to call it), has established an interesting precedent for the regulation of biotechnology.* Irradiation obviously had the regulators stumped, so they categorized it as an additive to be approved (and labelled) on a case-by-case basis. GE foods were not even on the horizon at the time.

In Canada, food irradiation was regulated as an additive under the Food and Drug Regulations and permitted for limited use after 1960, though in fact it was actually used commercially only briefly in the mid-1960s. In 1983 the Department of Health and Welfare proposed reclassifying it as a process, but this change was not implemented until 1989, at which time the Food and Drug Regulations were amended to include specific labelling requirements for irradiated food "to ensure the protection of consumers' rights to be informed and to enable them to choose between irradiated and non-irradiated foods."[145]

Only very slowly (which does not necessarily mean cautiously) have specific foods actually been approved for irradiation, but the relentless pressure of the nuclear industry and the growing presence of hostile pathogens in beef and poultry led to the approval of irradiation for both of these commodities in the US in December 1997.** Once this was achieved, the FDA changed the labelling requirements so that instead of having to display prominently the radura, the international irradiation symbol, manufacturers simply have to list irradiation among the ingredients, however small the type. Canada has not yet followed suit. There is still little irradiated food actually on the market, however. The processors are not eager to take it up because of its cost (10 to 15 cents per pound for meat), and the retailers don't want to touch it with a barge pole due to consumer antipathy. Articles promoting irradiation continue to appear in women's magazines and some industry trade journals nevertheless.

The recognition and certification of organically grown foods, both meat and fruits and vegetables, has created a second awkward precedent for regulatory agencies that now want to reverse the traditional

---

* Irradiation is a process of passing powerful X-rays, high-energy beams of electrons or radiation from radioactive cobalt, into food to kill microorganisms and prolong shelf-life. It can also be described as the subjection of food to radiation in the hopes of killing all signs of life, including any pathogens.

** Implementation, however, would have to await USDA approval of the necessary regulations, not expected until early in 1999.

approach in the case of biotechnology, and say that how food is produced or processed is irrelevant; it is only the "safety" of the product
that counts. In the case of organics, the whole issue is, on principle, one
of process. The only point at which organic certification might focus on
the product *per se*, apart from verification of authenticity by means of
a paper trail, would be by way of monitoring for possible toxic residues
acquired from unknown or external sources, such as spray drift from
nearby conventional farms, or organic bananas becoming contaminated when stored in the same sheds as heavily sprayed conventional
bananas.

The nuclear and biotech industries took notice of the problems that
official recognition of organic food production could cause them and,
in 1997-98, tried desperately to corrupt the proposed official US organic standards so that irradiation and genetic engineering would be
acceptable within the definition of organic. When the proposed standards were published, the immense public outrage at the inclusion of
genetic engineering and irradiation (more than 300,000 interventions
to the US Department of Agriculture in protest) forced the USDA to
withdraw the proposed standards and the Secretary of Agriculture to
apologize for the mess. Much more quietly, but at the same time,
Canadian organic certification bodies were voting on national organic
standards. The vote was resoundingly negative because somehow,
between the third draft that had been approved by the board working
on the standards for years and the fourth draft actually sent out, subtle
changes had occurred, such as the allowance of genetic engineering
within the definition of organic. Faced with overwhelming rejection, it
was explained as a "clerical error."

Since then the biotech industry, at least, seems to have decided that
it would be better off to stay out of that arena and allow organic standards specifically excluding genetic engineering to be implemented.
This will, of course, allow the biotech industry to claim that consumers
have the option to buy certified organic if they want to avoid GE foods
(as with potatoes) and that, therefore, there is no reason to label GE
foods! The biotech industry knows, at the same time, that there is
nowhere near sufficient production of organic foods to satisfy even
existing demand, much less to meet the needs of everyone who oppos-
es GE foods. The biotech industry also has to know that gene flow from

transgenic crops will be an increasing threat to organic production as the area of transgenic crop production increases due to a deliberate reduction in availability of good non-GE seed. Most of the seed companies in the world are now owned by one or another of the half-dozen largest biotech companies, namely Monsanto, Novartis, Dow, DuPont, AgrEvo, and Zeneca. In March 1999, DuPont bought Pioneer Hi-Bred, the world's largest seed company.

Irradiation and organic certification (along with kosher and halal identification, which also have to do with process — in these cases the process of slaughtering) provide curious precedents that it is the process, not the product, that matters. Like genetic engineering, of course, both irradiation and organic farming are production processes and the outcome of their use is products with subtle, but distinctive, characteristics.

To overcome this troublesome precedent, the Agriculture Canada office responsible for the regulation of biotechnology adopted a new category of "novel foods" for the products of genetic engineering, thus enabling the regulators simply to ignore the problem. A novel food is simply one with which Canadians are not familiar — which could apply to a tropical fruit not previously imported or a transgenic potato. The purpose of this category is to bury the distinctiveness of GE foods.

The strategy for achieving the desired changes in the regulatory regime — desired by the biotech industry, that is — has been pragmatic and piecemeal, that is to say, reductionist. On the grounds that we eat the product and not the process, the Canadian Food Inspection Agency (CFIA) says its responsibility is to regulate products, not processes. Canada also established a policy of evaluating/assessing (both terms are used by the CFIA) the products of biotechnology on a case-by-case basis. This conveniently avoids any principled decision that might raise a red flag and attract public attention. The gradual and quiet changes in the US regarding irradiation, such as extending the range of products which may be irradiated, illustrate how well this works.

A growing presence in food regulation is the global organization *Codex Alimentarius*, established in 1962 by the World Health Organization (WHO) and the Food and Agriculture Organization (FAO) of the United Nations "to guide and promote the elaboration and establishment of definitions and requirements for foods, to assist in their harmonization and, in doing so, to facilitate international trade."[146] It has been the job of Codex to establish definitions and standards for foods, food ingredients and additives, and food processes that could be adopted voluntarily by those countries wishing to use them in international trade. Food safety was not its responsibility.

Now, however, the United States, on behalf of the major corporations engaged in the global food trade, along with Canada, New Zealand, and Australia, is trying to force a change in the rules of the game so that *Codex Alimentarius* can become a compulsory standard-setting body for the purposes of the World Trade Organization (WTO). The desired outcome is that all national regulatory processes would have to conform to the WTO/Codex standards, which would be enforceable by the WTO. Cultural preferences, socio-economic and public health considerations, and any other "restrictions" that do not conform to the requirements of "sound science" would be prohibited on the grounds that they are irrelevant to food safety, thus giving free rein to companies such as Nestlé, Unilever, Monsanto, and Archer Daniels Midland to move foods and food components across borders — and down our throats — without hindrance.

At the same time, Codex is proceeding to recognize the production of organic foods as a process that can be labelled as such for international trade and which excludes genetic engineering, as indicated above. One intervention in the discussion by the International Seed Trade Federation/International Association of Plant Breeders was as pathetic as it was revealing: "The seed industry considers that the provision aimed at excluding GMOs [genetically modified organisms] from the scope of organic agriculture is absolutely inconsistent and not scientifically based .... If organic agriculture is not synonym [sic] of archaic agriculture, utilization of GMO varieties resistant to pests and diseases should be strongly encouraged in the guidelines."[147]

Contradictions

Protecting the public may still be given lip service by the conflict-ridden agencies that are supposed to be regulating "foods produced through biotechnology" (Codex terminology), but from the shifting role of *Codex Alimentarius* to the downsizing of food inspection agencies and the redefinition of what needs to be regulated and how, it is clear that the primary role of regulation is being shifted from protecting public health to market facilitation on behalf of the drug/biotechnology industry.

This is not the fault, necessarily, of the civil servants staffing the agencies, though in the case of the CFIA and Health Canada the alignment of the top officials dealing with biotechnology is and has been unequivocal. They faithfully express the dominant culture and reflect the absence of democratic accountability, however much they protest that they are simply doing "sound science" and not politics, and heedless of the judgements of their own working scientists.*

French sociologist Bruno Latour provides a playful commentary on the social drama of science:

> This Science, capital S, is *not* a description of what scientists do .... It is an ideology that never had any other use...than to offer a *substitute* for public discussion .... It has always been a political weapon to do away with the strenuous constraints of politics .... Because it was intended as a weapon, this conception of Science ... has only one use: as the command, "Keep your mouth shut!"

> The second meaning of "science" is the gaining of access, through experiments and calculations, to entities that do not at first have the same characters as humans ....

> Let Science No.2 be represented publicly in all its beautiful originality — that is, as what establishes new, unpredictable connections between humans and nonhumans, thus deeply modifying what the collective is made of .... Science No.2 ... redefines political order as that which brings together stars, prions, cows, heavens, and people, the task being to turn this collective into a "cosmos" instead of an "unruly shambles."[148]

Canada's biotechnology strategy

Canada's federal biotechnology strategy was first put forward as an industrial development strategy by Industry Canada in 1983. An indus-

---

* Health Canada has been embroiled for years in a festering controversy over the approval process for rbGH (and many other drugs). It came to a head in 1998 with the airing of the department's dirty linen in Senate and union hearings, with even media attention now focussing on the public disquiet and criticism rather than industry whining.

try-dominated National Biotechnology Advisory Committee (NBAC) was appointed at the same time. Since then, the regulation of biotechnology has been gradually cobbled together on an ad hoc basis in response to demands by business for new product approval. In 1998 the federal government decided it was time to renew Canada's biotechnology strategy — to put new wheels under the mechanisms of public funding for the biotech industry — and initiated a highly manipulated (and manipulative) "consultation" process to ratify the terms of policy dictated by the biotechnology industry.

The industry attitude is well expressed in this single sentence from the 1998 report of the NBAC:

> "Like all revolutions, biotechnology creates its own context that imposes adjustments on society."[149]

The implication that we are expected to observe technology's agenda and adjust our society accordingly is reinforced in the report's first chapter:

> Biotechnology is big, both as an economic driver and as an enhancer of Canadian quality of life. Biotechnology has momentum as a transforming new industry. Don't stop it .... Don't legislate; regulate through more easily adjustable guidelines.

There is a clear logic in the committee's recommendations. Regulations rather than legislation serve the interests of large companies such as Nestlé and Monsanto and Cargill[150] because the big corporations can easily afford to maintain lobbyists to encourage the bureaucrats to change regulations in favor of the corporations they represent out of range of public scrutiny. Legislation would, on the other hand, involve the risk of some public debate, if only in parliament. Consequently, Canada's policy approach to biotech regulation has been explicitly one of using existing legislation (however inappropriate) interpreted by guidelines and regulations crafted by bureaucrats and industry advisors, rather than creating new and appropriate legislation. Squeezing "novel" (i.e., genetically engineered) foods into existing legislation requires some obvious contortion — or distortion, such as that required to classify irradiation as a food additive. On the other hand, treating GE pesticides, such as Bt, as additives for regulatory purposes when inserted in crops such as corn and potatoes makes good sense, since that is exactly what they are, in spite of industry claims to the contrary. The biotech industry is adamantly opposing moves in this direction by US regulatory agencies.

While philosophically — and in the business press — big business may condemn all regulation as interference in the free market and advocate self-policing as a better mechanism to protect the public (no company, they say, wants to put a product on the market that will earn it a bad reputation or a lawsuit), a regulatory regime like Canada's actually contributes to the welfare of large corporations. These companies can not only afford to lobby and to meet regulatory requirements and standards, they can even call for more stringent regulations if it suits their interests. Companies with the financial resources can quietly call for more stringent and lengthier testing. They can call for higher — and more costly — equipment and building standards for research and production facilities, just to raise the ante for smaller companies. The big companies can then salvage the small ones by buying them out or, more importantly, gaining control of their technology. The bigger companies can recoup the additional costs simply by raising product prices.

Patent litigation can be used in the same way, particularly now that there is a drive to shift the burden of proof of infringement from the company claiming infringement to the party being blamed for infringing a patent. This means that if Monsanto claims a small company is infringing on a plant patent owned by Monsanto, the accused has to prove its innocence. Providing such proof can be a far more costly exercise than a small company can afford, producing the result that it gives up or sells out. Monsanto wins either way.

The Canadian Food Inspection Agency

The Canadian Food Inspection Agency (CFIA) was established in April 1997 to consolidate regulatory functions previously scattered through four ministries — agriculture, health, environment, and fisheries. The CFIA is responsible for the regulation of biotechnology in most regards except human health considerations, which remain under the jurisdiction of Health Canada. There are those, however, who believe that the creation of the CFIA was seen as a way of appearing to address the question of conflict of interest. The CFIA could claim to be clean, while the regulatory functions were taken away from a compromised Agriculture Canada. The public need not know that it was the same office under a new name. Shortly after the agency was established, I asked the head of the biotechnology office of the CFIA about public accountability. I was told that the CFIA has to report annually to the minister. This means, in effect, that there is no parliamentary or public oversight of the agency.

When the minister of agriculture tabled the legislation for the new agency, he stated clearly its contradictory mandate: "Consumer protection and the promotion of Canadian trade and commerce will be the agency's prime objectives." The career bureaucrat named to head the new agency described his challenge as being "to create more efficient and effective systems ... in a way that will not compromise food safety or impede industry's competitiveness in the marketplace."[151]

The Biotechnology Strategies and Coordination Office of Agriculture & Agri-Food Canada (AAFC) explained, in a document dated the same day that the CFIA was established, that although "some individuals have expressed concern in the past that having research and development and regulation in one department constituted a conflict of interest, these two functions ... were always carried out quite independently from one another .... While AAFC successfully managed this dual responsibility in the past, with the creation of the new CFIA ... the responsibility for regulating agricultural products (including agricultural products of biotechnology) now lies completely within CFIA."[152]

What was not reported was that the Biotechnology Strategies and Coordination Office was shifted, in toto, from AAFC to the CFIA: same people, same desks, same phone numbers, same attitudes, same culture. It is therefore unreasonable to assume that the new regulatory agency is any different in substance than the old promotional office that was responsible for developing, with industry help, regulatory policy for biotechnology. Or that there is any deeper understanding of what it is they are regulating.

Shortly after the new agency was established, it was discovered that farmers had purchased — some had even planted — a variety of transgenic, herbicide-tolerant canola that had not been registered and approved. The seed contained the wrong gene! The CFIA was quick to pat itself on the back and say, "See, the system works" — except that the error was not discovered by the regulatory process, which in fact had not noticed it, but by the seed company, apparently, although just who discovered the error has been kept a deep dark secret. In any case, a year later the CFIA introduced a regulatory change requiring the biotech/seed companies to swear that they know what they are doing and what genes are actually in the seeds they are selling.

"The agency has introduced changes to its varietal registration policy that requires companies registering transgenic varieties to file an affidavit certifying they have done molecular testing on its genetic makeup," said a senior official of the CFIA. "Plant breeders must also describe what testing procedure they have used. The change was deemed necessary after an emergency recall of Roundup Ready canola seed in the spring of 1997 .... We always assumed the breeders did know they had the right gene in their varieties."[153]

I have been unable to obtain an actual copy of the affidavit. The CFIA says that even the blank form is confidential business information.

Perhaps we could overlook one such regulatory failure, but the report of a second similar but unrelated event is cause for alarm. A 1997 field test in Sweden of canola genetically modified to be resistant to AgrEvo's glufosinate herbicide Basta contained two unauthorized lines in addition to the one line authorized by the Swedish government Board of Agriculture for testing. The seed had been produced by AgrEvo at their Canadian subsidiary in Saskatoon, hence under the oversight of Agriculture Canada. The mix-up was uncovered during analysis of test data by the Swedish seed company Svalof Weibull, not by the regulatory agency responsible for approvals at the time, Agriculture Canada.[154]

It is hard to know whether it is indifference, ignorance, or deliberate deceit that explains the casual and careless attitude of the CFIA. For example, another information bulletin dated the day the CFIA came into being speaks of work done through biotechnology "to improve the quality and nutritional value of food" and then describes how "scientists have developed tomato varieties with slower softening rates [meaning] they can spend more time ripening on the vine, becoming more flavorful and still survive handling and shipping."[155] The fact of the matter is that Calgene abandoned the Flavr Savr because it had neither the flavor nor the shipping qualities to make it in the market. Nor was it ever available in Canada, although approved by Agriculture

Canada, and no nutritional benefits were ever claimed by Calgene for the ill-fated tomato. The CFIA's effusive claims for the non-tomato seem somewhat inappropriate!

ॐ

It would be nice (and certainly healthier) if we could assume that the regulation of biotechnology was an objective matter of science within a framework of democratically arrived-at social values and goals. It would also be nice to be able to assume that there had been some kind of open or public process to establish at least minimum standards of health and safety in the public interest. It would then be a matter for government agencies, staffed by impartial and competent civil servants with technical expertise, to evaluate the products and processes put forward by persons or organizations seeking approval for commercialization of these processes or products in the light of the established standards. This would require the regulatory agencies, in addition to reviewing and assessing the data supplied by the applicant, to perform independent testing to verify or invalidate the data.

Unfortunately none of this is true for the regulation of biotechnology, at least not in Canada, in part for reasons already discussed. It does not seem to be the case in the US either. According to Suzanne Wuerthele, who has worked in a regional office of the-US Environmental Protection Agency (EPA) for more than thirteen years and is considered a national expert in toxicology and risk assessment, the EPA has an official position of "fostering" biotechnology. She says:

> There is no process — across all U.S. federal agencies — to evaluate the hazards of GE organisms, no formal risk assessment methodologies .... No science policies .... No conferences where scientific issues of GE are debated. No understanding of the full range of hazards from GE organisms. No discussion of or consultation with the public to determine what constitutes "unacceptable risk." No method to even measure magnitude of risks. Etc., etc.
>
> In the U.S., each risk assessment for GE organisms is done on an ad hoc basis by different scientists in different departments of different agencies. Some of these agencies have conflicting missions — to promote and to regulate; to consider "benefits"

as well as "risks." There is rarely any formal peer review. When peer review panels are put together, they are not necessarily unbiased. They can be filled with GE proponents or confined to questions which avoid the important issues, so that a pre-determined decision can be justified. These revelations and others have convinced me that this technology is being promoted, in the face of concerns by respectable scientists and in the face of data to the contrary, by the very agencies which are supposed to be protecting human health and the environment.[156]

Corporate determinism

Today the regulatory regimes of capitalist market economy states display a singular commitment to the expansion of their economy by pumping out new products, regardless of merit, with the least possible delay or interference. The production of goods and services is regarded as a sufficient national goal in itself. For their part, the regulatory agencies are expected to function as branches of the marketing operations of their corporate "clients." Senior staff of the CFIA, for example, frequently speak at industry events promoting biotechnology to reassure the industry that government is really on their side. The public is not usually invited to such events, and when it is, the registration fees are now set at such a level that only corporate-sponsored delegates can attend. This is not accidental, even though these events are almost always publicly subsidized.

To understand how we got to this point, we need to recall the cultural concept of technological determinism with its attendant erosion of democracy, which has been reinforced by a heavy propaganda campaign promoting market development and trade liberalization as the primary functions of national governments over the past decade and more. In this context we can more easily grasp the character of the regulatory processes as it has been shaped by this history, using the Canadian experience as an example.

One of the earliest policy documents on biotechnology in Canada was published by the Science Council of Canada, which for a few years in the early 1980s reflected the opposing views of uncritical acclaim of biotechnology by industry and critical evaluation from a public policy perspective. In 1980 the Council published *Biotechnology in Canada: Promises and Concerns*, and its title prefigures much of the industry and government literature on biotechnology. It is full of the promises of biotechnology, while the concerns of scientists and the public alike are treated as insubstantive fears.[157] At the time the Science Council was an arm's-length, government-financed, policy-analysis organization with some able members and a fairly respectable track record. (The Office of Technology Assessment in the US served a similar function and both met the same fate: when they became a nuisance to the biotech industry and other special interest groups, they were dissolved.)

The 1980 report expresses the ideology that shaped the development of biotechnology regulation: "Canada has a unique opportunity-to compete with other countries in the development of biotechnologies .... At this early stage ... we have the opportunity to incorporate social responsibility into a national industrial strategy." Unfortunately, the Council never defined this "social responsibility."

At the time there was at least one dissenting voice on the Council. Professor Stuart Ryan insisted that his critical comments be appended to the report. "As a result of my experience and examination of available evidence," he said, "I am more than concerned — I am frightened — when I contemplate the growing activity of the private sector in the field of biotechnology."

Ryan's voice can be heard in a 1982 report (referred to on page 139), but it does not seem to have affected the outlook of the Science Council for long, unfortunately. By 1985 a report by the staff, not the Council, clearly indicated that the Council was panting after biotech in time with the federal government: "Biotechnology may generate the last major technological revolution of the 20th century. The promise is already turning to profit; the pace is rapid; the potential is vast and exciting .... Canadians must grasp the opportunities offered by biotechnology if Canada is to improve its competitive position on world markets."[158]

Industrial strategy

When the first National Biotechnology Strategy was articulated in 1983 by Industry Canada, without benefit of parliamentary or public discussion, biotechnology was categorized as a new technology, the "bio" aspect considered as little more than a technological selling point having nothing to do with life. Although the strategy was the responsibility of Industry Canada, and the National Biotechnology Advisory Committee (NBAC) reported to the federal minister of Industry, Science & Technology, the development of regulatory policy was in the hands of a small office under the direction of Jean Hollebone in the Pesticide Directorate of Agriculture Canada.[159]

In 1987 the three regulatory departments — agriculture, environment, and health and welfare — agreed on several working principles, among them:

- to build on existing legislation
- to regulate the product as opposed to the process
- to build on internationally developed guidelines
- to use risk-assessment principles.

Hollebone subsequently explained that the regulation of biotechnology was "product-based" because federal legislation was primarily set up to deal with products, not processes. This approach allowed the government to build on existing legislation and did not involve "the expensive and slow process of building a new act," while allowing flexibility, such as establishing administrative guidelines to set requirements.[160]

Around 1993, Hollebone's office was recast as the Biotechnology Strategies and Coordination Office and in 1997, as already noted, it was put under the jurisdiction of the Canadian Food Inspection Agency. It is now referred to as simply the CFIA Office of Biotechnology.

Before Hollebone moved to another position in 1995, she oversaw the publication of the *Agricultural Biotechnology Regulatory Information Manual* (ABRIM). Weighing in at 3.75 kilograms, it provides the only comprehensive, though now dated, overview of the AAFC perspective, as well as providing virtually all of the information regarding the regulation of the processes and products of biotechnology under the purview of AAFC at that time. Nothing has replaced it, though individual documents are now available on the CFIA website. The problem for anyone wanting an overview is that the documents reflect the reductionist science they are based on, so one can view bits and pieces without ever getting a picture of the whole or an understanding of the real process behind the bits and pieces. The industry, however, understands how to work the system.

The unashamed bias of the government regulatory policy appeared again in the 1991 government publication, *BIO-TECH Regulations: A User's Guide*. "Biotechnological processes have been used to improve the quality of life for thousands of years .... The recent development of biotechnological techniques ... has created exciting new dimensions in our potential to improve human life."

In the same year the NBAC warned, "There is a clear and present threat that Canada will be left behind other countries .... A balance must be struck between regulation and promotion, equity and efficiency, protection of the public and the environment, as well as the furtherance of private interests and economic growth."[161]

There is no mention whatsoever of risks or dangers, possible unforeseen events or undesirable consequences, or the consideration of alternatives.

The ongoing process of policy formation, now including Health and Welfare Canada, was succinctly characterized in a deceptively informal document issued in 1992, simply headed, "Information Letter, Subject: Novel Foods and Food Processes."[162]

In this notice, the category of "novel foods" is definitively established as the mechanism for obscuring the real issues raised by GE foods — as if the introduction of Caribbean yams or New Zealand kiwi fruit was comparable to Bt corn or Roundup Ready soybeans. All of a sudden there is concern about the health characteristics of non-traditional foods and food processes being introduced into the Canadian marketplace, ostensibly by the growing number of non-European immigrants.

The Information Letter defines a "novel food" as "any food that has not been previously used to any significant degree for human consumption in Canada. This definition includes the use of existing foods for roles in which the food has not been previously used, and existing foods that are produced by a novel process." GE plants might — or might not — produce what the CFIA would consider novel foods. They are classed by the CFIA as PNTs — "plants with novel traits"* — which do not require regulation if they are considered "familiar" or

---

* "Plants with novel traits" are defined as plant varieties/genotypes possessing characteristics that demonstrate neither familiarity nor substantial equivalence to those present in a distinct, stable population of a cultivated species in Canada and that have been intention-

"substantially equivalent" to normal, non-GE counterparts. While the biotech industry and the regulators may have thought they had solved a major problem by adopting the category of "novel foods" (borrowed from the Organization for Economic Cooperation and Development, OECD, an agency established by the wealthy industrialized countries to promote trade and economic development), they also created yet another contradiction that may come back to haunt them. The principle of using existing legislation is based on the premise that there is nothing new or novel in the practice of biotechnology. Yet a corporation seeking regulatory approval of a transgenic plant on the grounds of familiarity may also be seeking a patent on the same plant — and one of the criteria for patent eligibility is novelty. The industry expects to have it both ways.

Labelling and liability

The labelling of GE foods ("foods produced through biotechnology") has been a bone of contention for years. As previously discussed, the issue first came up with rbGH more than a decade ago when the biotech industry realized that people might not be happy with strange hormones in their milk or GE foods in their shopping carts if they knew what they were and had the choice of whether or not to purchase and consume them.

In a sense labelling is not a regulatory issue, but at the same time the issue of whether and how products are identified has a long history; as long, in fact, as there has been trade over any distance at all. Without honest identification (and standard, verifiable units of weights and measures), trade, particularly international trade, would not get

---

ally selected, created, or introduced into a population of that species through a specific genetic change.

"Familiarity" is defined as the knowledge of the characteristics of a plant species and the experience with the use of that plant species in Canada. "Substantial equivalence" is defined as the equivalence of a novel trait within a particular plant species, in terms of its specific use and safety to the environment and human health, to those in that same species that are in use and generally considered as safe in Canada, based on valid scientific rationale.

According to Agriculture Canada Decision Document 96-06 on NatureMark potatoes, "The PNTs can either be derived from recombinant DNA technologies or from traditional plant breeding. Regulated field testing is necessary when the PNTs have traits of concern, i.e., the traits themselves, their presence in a particular plant species, or their use are: (1) considered unfamiliar when compared with products already in the market; (2) not considered substantially equivalent to similar, familiar plant types already in use, and regarded as safe."

very far. So it is strange indeed to find companies that are major proponents of globalization, harmonization, and open markets arguing that their goods should not be adequately and honestly labelled. They discuss with great seriousness the wants, as distinct from the needs, of the public and argue that what the public needs to know will be best decided by the manufacturers and regulators, regardless of what the public wants to know or feels the need to know. The industry expects the regulatory agencies to uphold its illogical and paternalistic arguments.

Perhaps with this in mind, *Codex Alimentarius* first adopted a Code of Ethics for International Trade in Foods in 1979. "The objective of the code is to establish standards of ethical conduct for all those engaged in international trade in food or responsible for regulating it."[163]

The General Principles of the Code state, among other things, that "International trade in food should be conducted on the principle that all consumers are entitled to safe, sound and wholesome food and to protection from unfair trade practices," and "No food should be in international trade which: ... is labelled, or presented in a manner that is false, misleading or deceptive."[164]

The trading and marketing of foods "produced through biotechnology" and not labelled as such is clearly deceptive and unethical by the standards of Codex itself, but the debate within Codex about the actual labelling of biotech foods has been going on for several years and shows no signs of coming to an easy conclusion.*

Obviously the issue of labelling "foods produced through biotechnology" is far bigger than simply deciding whether or not a label or identification should be on every food produced through biotechnology (that is, transgenic or produced from transgenic plants/crops) at the point of retail sale. Among the more complex questions lurking behind the label are those of liability and burden of proof. In addition, any discussion of labelling has to recognize more general socio-democratic questions such as: What do you need to know? Why do you need to know it? How should you get to know it?

These questions are particularly important when the regulatory agencies have administrative procedures based on principles that sub-

---

* Canada is the host country for the Codex Committee on Food Labeling, which meets annually in Ottawa.

tly offer answers many of us would not agree with. One such principle is that of risk assessment itself, which includes the determination of what constitutes an acceptable risk. This raises the obvious question of who should make this decision, and on what grounds.

The Science Council of Canada recognized this in an exceptional report published in 1982 titled "Regulating the Regulators." "Government departments involved in value-scientific disputes should make explicit their consideration of alternative policies, the likely areas of future concern, the range of scientific uncertainty and the probabilities of anticipated risk."[165]

This advice was obviously ignored in the development of regulatory policy, allowing the CFIA to reduce risk assessment to a technical issue: "Risk assessment, conducted by evaluators at the Canadian Food Inspection Agency (CFIA), is the process in which each risk is identified and reviewed in the light of the scientific information available. 'Safety' does not imply the absence of risk, but rather a level of acceptable risk...The level of risk is determined to be acceptable if the new product is as safe as its traditional counterpart."[166]

The doctrine of "acceptable risk" relieves the CFIA of responsibility and liability because the standards of "acceptable risk," "substantial equivalence," "familiarity," and "sound science" are sufficiently subjective that liability or responsibility cannot be established. It then comes down to *caveat emptor*, or "buyer beware."

"Passing the buck" is another colloquial term for this. Legally it is called "burden of proof," as in "shifting the burden of proof." An example of this is found in industry discussions that interpret the absence of evidence of "catastrophic" consequences, as Dale Bauman put it, as the evidence of absence of any unwanted consequences. The attitude is described by scientists and biotechnology critics Mae-Wan Ho and Ricarda Steinbrecher as "don't need — don't look — don't see." This effectively gives scientists *"carte blanche* to do as they please, while serving to diffuse and allay legitimate public fears and opposition."[167] As elaborated in the newsletter of the Science and Environmental Health Network:

> Risk assessment has been narrowed to a search for proof that death, cancer or some other adverse effect is caused by a product or technology. The key words are "proof" and "causality" ....

> Lack of scientific proof of harm does not mean a product or technology is safe .... In the face of scientific uncertainty, deciding whether a chemical or technology is safe becomes a matter of policy, not science. In those circumstances, risk assessment may be misleading. It may suggest that lack of evidence of harm is evidence of safety. It may give the benefit of the doubt to products and processes that are harmful.[168]

There is a clear policy alternative to the subjective and inadequate process of risk assessment, and that is the "precautionary principle," which puts the burden of proof back where it belongs: on the proponents of "plants with novel traits," "novel foods," and all other forms of biotechnology that induce changes we barely comprehend. The precautionary principle, as defined by the 1990 Bergen Ministerial Declaration, says: "Where there are threats of serious or irreversible damage, lack of full scientific certainty should not be used as a reason for postponing measures to prevent environmental degradation."[169] In other words, err on the side of caution. "Precautionary science invites us to make explicit the boundaries of our knowledge by unearthing complexity, ignorance and values ... thereby revealing how our concepts of certainty are defined."[170]

Rather than protecting the public from the special interests of corporations, the governments of the western capitalist countries by and large take it as their responsibility to protect the corporate sector from the public. Government regulation is actually desired by the biotech industry, not only for the reasons stated earlier, but also because approval by an agency such as the CFIA can be used as a seal of "Good Housekeeping" to protect the industry from public criticism. The Food and Consumer Products Manufacturers Association of Canada, for example, says, "One of government's roles is to communicate to consumers about Canada's excellent regulatory system that ensures one of the safest food supplies in the world."[171]

This protection of the corporation (provided it is big enough) is a disturbing characteristic of the culture of globalization, harmonization, and corporate welfare. It turns up everywhere. There is a process underway in the field of intellectual property rights, for example, to shift the burden of proof in claims of patent infringement from the patent holder to the party accused of infringing a patent. Ian Wilmut and his colleagues at the Roslin Institute claimed a patent on the technique by which they succeeded in cloning Dolly the sheep. Subsequently, a team in Massachusetts claimed to have cloned cows using a different technique. Wilmut accused them of infringing on his patent, feeling that it is up to the Massachusetts team to prove that they have not. Shifting the burden of proof can also be a simple matter of intimidation if a company such as Novartis accuses me, or my company, of infringing on a patent it holds and I have to incur the expenses of proving I have not. This provides a wonderful stick with which big companies can beat little ones, regardless of the actual legal outcome.

Another unspoken change in the concept of burden of proof is evident in a half-hour training video for Agriculture Canada inspectors. The video describes biotechnology as the manipulation of genetic material "to provide goods and services," and situates genetic engineering in the familiar continuum of brewing and bread making, saying that it "holds much promise" for all kinds of wonders. It also subtly shifts the burden of proof with the narrative statement that claims it is a "belief among experts" (unidentified) that GE products are "not significantly different" than their traditional counterparts. This is a telltale warning that the proponents of these "plants with novel traits" will no longer have to even show substantial equivalence; it will be up to the regulator to show substantial difference. The video had no introduction or credits at the beginning, but at the end across the screen in big letters was "We wish to thank: Monsanto Canada." The next frame said, "This training film was produced for the use of the Food Production and Inspection Branch of Agriculture Canada."[172]

The conclusion of this, of course, is in the stomach, and at this end, as I have already mentioned, the principle being put forward is that if I get sick from eating genetically engineered Bt potatoes, the burden of proving what did me in is up to me. Monsanto claims it is not responsible because the potatoes were approved by the CFIA. The CFIA can pass responsibility back onto me because the potatoes met their (subjective) standards. Who gave the CFIA this authority? The act of parliament establishing the agency. Can anyone actually be held responsible for my getting sick (or dying) from eating an approved, but unlabelled, GE food? No, because the whole regulatory construct rests on the foundation of "sound science," and science, after all, cannot be wrong.

Besides which, biotechnology is essential if we are to feed the world and save the environment!

More common logic would suggest a scenario in which the corporations engaged in genetic engineering would accept, and be held to, full responsibility for what they are doing, since theirs is a profit-seeking venture, not a public-interest project. The corporations would insist on labelling and would charge a premium price for their "improved" products — not just at the farm level for seeds, as at present, but right through to the retail checkout.

Liability would fall on the corporations (as it does on medical practitioners), and when submitting a product or process for approval, the burden of proof would rest on the applicant to prove safety, not on the regulator to prove harm. As Hiltrud Breyer, member of the European Parliament for the German Greens, put it, "If researchers are so sure that their genetic experiments are safe, why are they afraid to accept liability? How does the biotech industry expect to earn the confidence of the public if it refuses to take public responsibility for its actions?"

The regulatory agencies would not thereby be relieved of responsibility, but they would be held to strict, objective standards. If, for example, a food was produced through biotechnology, the product would be evaluated more comprehensively than as a novel food, and labelling would be required since so little is actually known about the long-term consequences of biotechnology, including human health effects. This would at least enable any subsequent problems to be traced to their source.

The regulatory process would then be required to be fully open to public scrutiny, as recommended by the Science Council in 1982. If a company sees benefit in marketing a new product, it has to be willing to expose it to public examination, not claim secrecy on the grounds of proprietary interests, the argument that both corporations and government currently use to maintain a cloak of secrecy — confidentiality they call it — around the regulatory process.*

---

* This would be a far cry from the current situation described by Mae-Wan Ho and Ricarda Steinbrecher in an analysis of a UN consultation report that, they say, shows how the regulation of GE foods is really designed to expedite product approval "while serving to diffuse and allay legitimate public fears and opposition."

   The principle of substantial equivalence (SE) ... is completely unscientific and arbitrary. ... The principle is not only vague and ill-defined, it is flexible, malleable and open to interpretation. "Substantial equivalence" does not mean an equivalence of the

The regulatory agencies could accept responsibility for what they are regulating because of the open and objective character of the process, and approved products and processes would carry labels adequate to establish responsibility, i.e., Bt potatoes would be labelled as such and their source would be indicated so that they could be avoided or, if consumed with deleterious effects, the victim might know what caused the reaction and hence what to do about it. The regulations governing the production and distribution of GE foods would be based on much more objective and conservative criteria, on the grounds that the purpose of the regulatory process is actually "to assess new products for efficacy and safety in order to protect humans, animals and the environment" and not to "assist Canadian companies in maintaining the quality and effectiveness of products that are traded internationally." (These contradictory purposes are contained in the second paragraph of the CFIA document referred to above.)

Obviously labelling is all mixed up with the product/process debate. Industry and the regulators say that a potato is a potato; the fact that it underwent a process of genetic engineering and came out a transgenic makes no difference to the eater, and therefore the eater does not need to know what happened to the potato. But as we have already seen, it does indeed make a difference to some people.

Finally, labelling is an important social issue because it is one gateway into positive choices for a different kind of society. The labelling of certified organic produce, for example, is both positive and negative: negative in that it says biotechnology and agrotoxins have not been used in the production of the food; positive in that it goes beyond the narrow risk/benefit health issues for the consumer of the product to

unengineered plant or animal variety. The GE food could be compared to any and all varieties within the species. It could have the worst characteristics of all the varieties and still be considered SE. A GE product could even be compared to a product from a totally unrelated species. Worse still, there are no defined tests that products have to go through to establish substantial equivalence ....

The Consultation explicitly failed to assume responsibility for ... the possibility of new viruses being generated and of genes jumping (horizontally) across species barriers, as the result of GE biotechnology itself .... The Report is openly partisan to the technology, making unsubstantiated claims for its benefits while omitting to mention the socioeconomic impacts on small farmers, and the viable alternatives to the technology in all forms of sustainable agriculture already practised worldwide.

(Mae-Wan Ho and Ricarda Steinbrecher, *Fatal Flaws in Food Safety Assessment: Critique of the Joint FAO/WHO Biotechnology and Food Safety Report* [1996], Third World Network, Penang, 1998.

the larger questions of sustainability, stewardship of land, and care of animals, farmers, and farm workers. Organic agriculture is also a social phenomenon. Not only is it generally labor intensive, it also tends to bring farmer and eater into direct contact, as in community shared agriculture (CSA). Sustainable/organic/ecological agriculture has a strong tendency to view the public not as consumers in search of cheap food, but as partners in a social enterprise to satisfy the needs of all.

Sowing confusion

> The objective of modern biotechnologies is to improve plant varieties by the introduction of new genes. The word "altering," which can be considered as a negative interpretation of biotechnology, is not appropriate. "Improving" would be wiser. However, in order to be neutral in the definition, we suggest to replace "altering" by "modifying." —International Seed Trade Federation/International Association of Plant Breeders[173]

In 1997 I participated in a seminar for high school teachers from across Canada, this one on the pros and cons of labelling GE foods. The North American PR official for AgrEvo, one of the big six agrotoxin/biotech companies, presented a detailed and absurd explanation to the teachers as to why it was impossible to segregate crops from GE seed and crops from conventional seed. Since segregation was impossible, she concluded, so was labelling. Then, to my amazement, she said, "Besides which, once the horse is gone it is too late to fix the stable door!"

Being next on the panel, I could not help saying, "Thank you very much, Margaret [Gadsby]. I am glad that Ray [Mowling, Monsanto's vice president] is also here, since you have just confirmed what I have long suspected, and if I am wrong you can tell me. What I have suspected is that you have an industry policy of getting as many GE crops through the regulatory process and on to the market — and into the supermarket — as fast as possible. I have also suspected that you have been pushing the deliberate mixing of GE and non-GE so that you can say, as you just did, it's too late to segregate and label."

There was no comment.

Clearly it is no accident that it is hard for the average citizen to distinguish science from science fiction — or sound science from nonsense. The very same companies that say biotech crops cannot be segregated are also engaged in developing genetically engineered specialty crops whose value lies in the fact that their identity is preserved from seed through to delivery to the processor (hence the label "identity preserved," or IP). Monsanto's GE high-laurate canola, which produces an oil that can replace palm oil in manufacturing detergents, is a good example.

IP crops need not be genetically engineered — they can be wheats with particular baking or milling characteristics, for example - but as public resistance to GE foods grows, so will the importance of strict segregation of the crop at every stage. Interestingly, some of the companies that are handling GE crops and claiming they cannot be segregated are also handling certified organic crops, which have to be kept completely segregated — and identified, i.e., labelled. It is obviously a matter of will, not of science or technology.

There is obviously a deliberate sowing of confusion: not only confusion of traditional and transgenic crops, and images of novelty with images of familiarity, but the genetic confusion of the crops themselves. If the growing number of scientist critics of biotechnology are right, no one really knows what is being sown in the fields of transgenic canola, potatoes, corn, cotton, or soybeans. Look at it from the crop's point of view. The transgenic Bt potato must be wondering what it is supposed to be producing: Bt for the beetles or tubers for the dinner table? The Roundup Ready canola must wonder if it is still an oilseed maker or just a herbicide factory? As biologist Sonja Schmitz of the University of Vermont puts it, the challenge to the plant is just like an arms race, where a single project commands all the resources at the expense of all other responsibilities and projects.

Not surprisingly, this genetic confusion is beginning to be suspected as the cause of diminished vigour and performance in transgenic crops such as canola.

> Yields of the [Innovator] canola, which is tolerant to Liberty herbicide, did not materialize and germination was patchy .... Saskatchewan Wheat Pool compensated about 90 farmers who had complained .... Claims by another 75 people will not be honored. The seed was developed by Agriculture Canada

and AgrEvo.[174]

> Breeding programs might ... be pushing canola too hard. While trying to increase yields of oil and protein, they may be decreasing the seed coat to the point where chemical treatments are getting into the seed and damaging the embryo, according to Tony Zatylny of the Canola Council of Canada.[175]

> There appears to be a growing problem with seed that germinates but does not develop into a viable crop.[176]

Studies on the costs of introducing herbicide resistance into transgenic crucifers have measured seed production and the ability to overwinter — important aspects of plant survival. They report that transgenic plants were less fit than their nontransgenic counterparts, and in the absence of herbicide selection pressure would not be expected to survive over the long term as well as nontransgenic plants. The reason for the decreased fitness was not clear.[177]

An obvious part of the problem is the corporate race to put more and more new varieties on the market every year. In 1982 there were six new canola cultivars, all developed in Canada through public breeding programs. A single new variety was being released each year after thorough comparison trials. To be registered, a new variety had to be as good as or better than the check variety (an established well-known variety). With the push to get new "products" on the market — an ever increasing number of "improved" varieties — the government allowed the seed trade to eliminate the long-established cooperative trial requirement, in which varieties were grown side by side to determine how different varieties perform in real life under the same conditions. The federal government also changed its own rules for seed certification, which are based on a merit point system. Now a new variety is awarded extra points simply because it was genetically engineered. The result was that by 1996 more than 130 varieties were on the market, with 30 new ones being released each year, more than half of them transgenic, and each having a life span of about three years before being displaced by yet another "new" variety.

As harvest got underway in 1998, I noticed a telling ad in *Western Producer*, a major farm newspaper of the prairie provinces. The ad was headed: "ATTENTION SMART CANOLA GROWERS — JUST HOW

SMART ARE WE?" "Smart" is the brand name for one of the herbicide-tolerant, but not technically transgenic canolas on the market. The ad read: "If you are having problems with Smart Canola seed and are not satisfied with the settlement being offered to you by Proven Seeds,* we would like to hear about it. Perhaps if Proven Seeds is not listening to us as individual farmers, they may want to PAY attention to a group of us. Please send your information and concerns to...Neepawa, Manitoba."[178]

The biotech industry now faces a dilemma of its own making. On the one hand it has to keep the public at bay — or at least docile — while at the same time it has to generate the hype that will continue to suck investors' money into the black hole of biotech. As *New Scientist* expressed it, "Fuelled by hopes and dreams rather than actual products, [European biotechnology companies] have been racing to raise funds with all the reflection of fortune hunters in a gold rush."[179] "Observers ... claim the [biotech] industry ... made promises and predictions to investors that are proving difficult to fulfill...Of the 430 biotech companies listed on stock exchanges across the world, only around 10% are profitable."[180]

One way of maintaining investor confidence is to hold out the bait of a pipeline (Monsanto uses this term) chock full of blockbuster drugs and transgenic crops, all of which will, of course, gain prompt regulatory approval. It is a world of promises, as biotechnology has been from its conception. As Robert Bud points out, more than two decades ago considerable efforts were already being made "to reduce media anxiety and to emphasize the potential practical benefits" of the new biotechnology. "Influential formulations of the potential were expressed and generalized to defend the field from regulation. Uses of recombinant DNA technology were portrayed long before most were practical, to ensure that industrial development wouldn't be impeded .... As early as 1974, a list of speculative gains was being set against the speculative risks."[181]

The research scientists, of course, get recruited as stock promoters and drug salesmen, leading Harvard biology professor emerita Ruth Hubbard to write, "I do not mean to suggest that molecular biologists

---

* Proven is the brand name for seeds distributed by United Grain Growers, a publicly traded cooperative that is 49 percent owned by Archer Daniels Midland of the US.

are deliberately deceiving people when they advertise the potential effectiveness of their work. Some may be doing that, but the more significant point is that they are members of this culture, which is ready to devote huge sums and much effort to eliminating biological causes of illness and death while at the same time accepting as inevitable a steady increase in the death toll from social causes."[182]

Hubbard points out that by focussing our attention on microorganisms or genes, corporate scientists draw our attention away from social influences and ensure their own monopoly. Health gets defined as a scientific problem for which we must seek scientific answers. "[But] an excessive preoccupation with individual concerns and responsibilities is detrimental to health when it encourages us, as a society, to neglect the systemic conditions that affect us all."[183] "Attention and funding is diverted from the social to the molecular," as Steven Rose puts it.[184]

There is also the practical aspect, as Hubbard says, that "pharmaceutical companies and physicians stand to make a good deal of money from inventing new diseases as fast as new diagnostic tools are developed that can spot or predict their occurrence."[185]

A cultural preoccupation with personal health, perfection, and longevity plays right into the business interests of the "life sciences" corporations, which are happy to redeem our bodies from death, one piece or one gene at a time.

Lobbying — Making sure it happens

I mentioned earlier that from its very beginning, genetic engineering in all its forms and manifestations has been, not by accident, categorized as technology. In a culture of technological determinism, the assumption that technology is an autonomous force conveniently removes it from the realm, and possibility, of democratic control. It also reduces corporate executives and government officials alike to the function of technological agents, free of any moral responsibility for the consequences of the production or approval of new technologies, including the products of biotechnology. The executives' and officials' responsibility is instead to get new products on the market to improve shareholder values and the national economic indicators.

The public message of the biotechnology industry, or "community," as it likes to refer to itself, is "progress through technological determinism." The actual functioning of the industry makes it quite clear that it prefers to determine "progress" its way, and it has little difficulty attracting the public funds necessary to do so. To take advantage of government funding, the industry has developed a penchant for creating lobby and PR organizations just as it creates plants with novel traits.

The premier biotech industry lobby group in Canada is Ag-West Biotech Inc., the first really aggressive biotech promotion organization in the country. Others, such as the Canadian Institute of Biotechnology (now BioteCanada), have been more traditional and restrained — some might say covert — in their lobbying and PR efforts. Ag-West was established by ex-DowElanco employee Murray McLaughlin in Saskatoon in 1989 with a $900,000 grant from the Saskatchewan provincial government.

The first issue of Ag-West's newsletter, *AgBiotech Bulletin*, wasn't published until 1993, but it soon established itself as the ideological and strategic voice of agricultural biotechnology in Canada. In the second issue, McLaughlin set out his criteria for public relations:

> There are two messages which we can deliver to the public to help them come to terms with the positive contributions of biotechnology. [First] the foundation techniques of agbiotechnology have been used for millennia ... [and second], the development of agbiotech is the result of a collaborative alliance between government, laboratories, university researchers and industry. Although profits are expected from eventual commercialization, the primary motivation for scientific development is improved quality of life for the community.[186]

For a number of years one could find ICAST — the International Centre for Agricultural Science & Technology — sharing offices and staff with Ag-West Biotech. ICAST was established as a not-for-profit corporation with a $10.5 million fund contributed by the federal and provincial governments. Management of ICAST was contracted out to Ag-West Biotech, with Murray McLaughlin as executive director. ICAST was quietly reabsorbed by Ag-West Biotech after McLaughlin left.

Past *AgBiotech Bulletins* and "Infosource" sheets (prepared by Ag-West for students) provide a litany of ideological aphorisms:

Farmers have always looked for new and better ways to control insect pests.

Helping to improve upon nature ...

Science must continue to be the basis of regulation .... Other issues, such as socioeconomic and ethical issues, are too variable and could be used by industry opponents to hold up the approval of new products indefinitely.

Scientists need to think about what words they use to explain their science. Manipulation, for example, can become modification ...

By harnessing nature's methods of gene transfer, scientists have been able to genetically alter a number of common crops, creating new varieties that are better suited to farmers' needs.

Our communication efforts are ... critical. The public has not yet fully accepted agbiotech.

After six years at the helm, in 1996 McLaughlin resigned as president of Ag-West Biotech to become deputy minister of agriculture in the Saskatchewan provincial government. A year later the Saskatchewan government, through McLaughlin's department, committed $1.1 million for each of the next four years to the work of Ag-West Biotech, while another provincial fund was to provide an additional $780,000 for the next three years "for the development of scientific programs related to public education and awareness and enhanced access to the regulatory process for Saskatchewan agbiotech companies."[187]

His mission completed in Saskatchewan, McLaughlin moved on to become the first president of Ontario AgriFood Technologies (OAFT), a not-for-profit consortium with offices in the new Ontario Ministry of Agriculture and Rural Affairs (OMAFRA) building in Guelph, Ontario. The mission of OAFT is to "provide leadership and coordination in utilizing biotechnology to generate wealth for the agriculture and food industries of Ontario." Its motto: "From Discovery to Profit."

The federal government committed $650,000 over three years to get the project off the ground, and OAFT is expected to have an annual budget of about $300,000. The consortium consists of five Ontario universities, the Ontario ministry of agriculture, most of the industrial commodity organizations (corn producers, soybean growers, etc.), and corporations such as AgrEvo, Cyanamid, DuPont, Monsanto, Novartis, Pioneer Hi-Bred, and the Royal Bank.

In January 1999 McLaughlin improved his position by moving to the Royal Bank, where he is responsible for assembling venture capital for biotechnology research projects.

Agriculture Canada has not only actively propagandized the public on behalf of the biotech industry since 1983, it has also provided in-kind and moral support for its lobby groups, most notably the Food Biotechnology Communications Network (FBCN). The FBCN was established in 1993 by seven corporations plus Ag-West Biotech and ICAST. Among its other goals, the FBCN aims to "encourage consumer confidence and enhance trust in the regulatory system" and "provide educational grounding which will help consumers understand the industry...and recognize the benefits it offers."

Today the FBCN describes itself as "the only national organization to bring together all the interests of food biotechnology, right from the farmer to the consumer." It also claims to bring "both neutrality and strong credibility to the information we provide."

In 1997 the organization moved its offices from Ottawa to the OMAFRA building in Guelph, where it set up shop right next to Murray McLaughlin and OAFT. Of the FBCN's 1997-98 budget of $153,000, $150,000 came from 60 corporate memberships — including AgrEvo, Dow, Monsanto, Novartis, and Zeneca — and $3000 from 100 individual memberships to make it appear democratic. In addition, AAFC contributed $120,000 for total revenue of $273,000. $145,000 of this was budgeted for what might be called evangelization. In its April 1997 newsletter, FBCN assured its members that "the messages from the special interest groups [promoting public resistance to GE foods] are countered with balanced, science-based fact."

Among the strongest supporters of the FBCN has been the Consumers Association of Canada, which has, strangely, been the industry's key lobbyist against the labelling of foods produced through biotechnology — "food labels are not the only way and may not be the best way for consumers to get the information to make informed decisions." CAC has joined with the FBCN to publish a gushingly biased "information kit" on the glories of biotechnology and the merits of GE foods. On the last page of the kit amidst a list of "issues" is a cute disparagement of public concern: "For some people the heart of the debate is in the morality of introducing new technologies." This is typical of industry tactics to avoid discussion of substantive issues and sincere public concerns. Morality is certainly the issue, but it goes far deeper than simply the introduction of new and unnamed technologies.

Several years ago the Biotechnology Coordination and Strategies Office of Agriculture Canada published a high-gloss promotional piece for the products and processes it is supposed to be regulating. The booklet, which is still in circulation, lists examples of the possibilities of biotechnology: "improved crops ... enhanced food products ... better, healthier animals."[188]

The most extreme example of hyperbole and misinformation I have seen, however, is a booklet for classroom use, produced by the Canada-Saskatchewan Agriculture Green Plan Communications Committee and paid for by the provincial and federal governments in 1996 "to provide the public with greater awareness and understanding." An example of the text: "All tools and techniques of biotechnology share these common features: the technology is highly refined; it is operated with a very high degree of precision; the technology is used in controlled environments." The booklet does not explain how several million acres of transgenic Roundup Ready canola is grown in a "controlled environment," or how gene drift operates with "a very high degree of precision."

Along the way, Agriculture Canada has been redesigning itself to better serve the corporate sector. In the mid-1980s Eugene Whelan, Liberal agriculture minister, allowed the name of his ministry, Agriculture Canada, to mutate to Agriculture and Agri-Food, apparently a misreading of Agriculture and Agri-Business. The research branch of AAFC gave itself the mission of moving "from basic science toward researching what works best to command agrifood markets. It is here and now research seeking specific results in each and every project, with short to medium-term payoffs."[189] To push this along, the "matching investment initiative" of the federal government provided a dollar's worth of AgCanada science for each dollar of farmer and industry money spent on specific projects. The shift in focus, said the head of research strategies and planning for AgCanada, gives industry clients "much more say in what we are doing. Simply by putting a dollar on the table, you are setting priorities."[190] A 1998 government of Canada brochure, "Investing in Canada's Dynamic Agricultural Biotechnology Sector," advertised the "financial benefits" of tax credits and direct financial support, saying, "Investors look for the highest returns and the shortest payback periods on their ag-biotech research investments." The government was stating clearly where its priorities lay, and they did not include the well-being of Canadians.

A partnership between Dow Chemical subsidiary Dow Agrosciences and Guelph-based Performance Plants (an offshoot of the University of Guelph) illustrated this program in action. At a press conference announcing the venture, Arthur Carty, president of the National Research Council of Canada, described it as a marriage between entrepreneurial science, business and long-term government funding. Corporate speakers agreed, heaping praise on Canadian government support.[191]

Which puts policy squarely in the hands of "the capital community," as the ad for the National Research Council of Canada Biotechnology Group expressed it.[*]

It would be wrong, however, to hold the biotech industry and its government cronies alone responsible for this state of affairs. The industry, after all, is only pursuing its own self-interest. The government employees, for their part, are intelligently, if immorally, considering their own future employment in an age of privatization. The revolving door between industry and government means that government and industry employees are frequently on the move from one to the other. It contributes to understanding and good communication.

The whole sorry state of affairs comes back to an issue of democracy and the requirement that the public take responsibility for technology and regulatory policy. This means the public must create the struc-

---

[*] "The activities of the Biotechnology Group have been translated over the years into enhanced industrial activity and investments in biotechnology by the capital community throughout Canada." (National Research Council ad in *Biotech*, the magazine of The Biotechnology Network, Fall, 1998)

tures for debating and forming public policy democratically, and for holding the government accountable — to the public rather than to just a diminishing handful of transnational corporations.

Chapter Ten

Lifelines

The organism is both the weaver and the pattern it weaves, the choreographer and the dance that is danced. — Steven Rose[192]

Perspective and context

Having now looked in detail at biotechnology in practice, we need to pick up the larger story from where we left off in Chapter Five.

The social construct of modern industrial biotechnology is based in the western cosmology of linear history and progress. Linearity itself has two directions of equal value, like a railroad track, but progress has to be both linear and unidirectional by definition. The arrow of progress carries us to our destination, into the future, for better or for worse. We are just along for the ride.

In this determinist context, technology is both a means and an expression of progress. It is both the process (or collection of processes) by which we achieve progress and, in turn, a product of the progress. Whatever is classed as "technology" needs no explanation or justification in this system; it simply is, and we are expected to give it due respect and allow it to carry us forward. We are not to ask who classed it as technology or why.

This somewhat mindless cosmology finds expression in comments such as, "Which road cloning will go down is still not clear. All we know now is that there is a road," in the words of a *Toronto Globe & Mail* editorial.[193]

A culture of linearity and progress is not inclined to make space for differences of opinion — or for other perspectives or cultures — particularly when it is combined with the kind of triumphant universalism that marked the end of the Cold War and the defeat of the Soviet Union. Unfortunately, the record of how such a civilization deals with deviants —from throwing the outcasts to the lions in a public spectacle, burning them at the stake in the town square, or threatening them with lawsuits or loss of job — is not attractive. The history of biotechnology reflects this monocultural approach: no dissent is heard or recognized.

If history is the arrow of progress and no deviation is possible, then the only issue is, How fast? Perhaps this is why speed is good and faster is better in the collective mind of the biotech industry. It may also explain why the regulation of biotechnology has taken the subservient form of getting more products to market faster, as described in previous chapters.

As an expression of this culture, the language of the biotechnology industry is riddled with other subjective, ideological, and unscientific words, words such as "improved," "better," "forwards," and "faster." One also finds words such as "backwards" used frequently to describe anyone who dares question the progress of biotechnology. The implication is that we can move only forward or backward, but never sideways — and it is immoral to go backward. Even if I am standing at the edge of a precipice in a fog, I am morally obliged to move along, forward. In reality, of course, if I find myself in that situation, or facing a wall, I know perfectly well that I have a number of options: I can, indeed, walk into the wall or over the cliff; I can stop; I can back up or even turn around, if only to get a better perspective — or gain an insight — on my situation and to consider alternatives, which might mean moving sideways, in either direction, or even going back home. In other words, I can confine my life — and consciousness — to the narrow dimensions and single trajectory of "progress," or I can recognize the contextuality and complexity of my life.

Robert Horsch, a senior Monsanto executive, illustrated the industry use of the logic — and presumed moral weight — of relentless progress in a 1997 speech that is still highlighted on Monsanto's website as a corporate policy statement. Monsanto drew on this logic in an expensive bid to win public support for its transgenic products in the United Kingdom and Europe during 1998.

> We can not afford to squander our time debating that last fraction of risk from fantastic "what if" scenarios of biotechnology that ignore 10 millennia of broad experience in agriculture and tremendous modern insights into the science of genetics.

> The freight train bearing down on us is the crisis of sustainability and sustainable development. The precautionary principle tells us that even without full certainty about the paths ahead, we should act to avert the serious and irreversible harm that is occurring even as we speak. Failure to move forward with new technology ... global trade and business development ... is probably the biggest risk we face. Our inertia on our historical track will kill us for sure if we act too slowly.[194]

Here is a clever reversal of images and issues, as well as the inversion of the precautionary principle, which says, as we described earlier, if you don't know what you are doing and what might happen as a result, particularly if the consequences could be negative or disastrous for people who never asked you to do it, then you should not do it.

The traditional identification of technology with the train coming down the tracks, or down the road, is reversed in the demagoguery: it is ecological disaster that is coming down the tracks if we do not abandon the precautionary principle and embrace this salvific technology of genetic engineering (our only hope of salvation) with all our strength and devotion.

If we reject such moral blackmail, however, and step aside, off the tracks and into a little clearing, or climb a nearby tree, and let that technology carry on without us, what might we observe and feel? What power can that technology exercise over us if we choose to get out of the way? The train of technology — or its corporate driver— cannot so easily leave the tracks and pursue us.

&

Barbara McClintock was a pioneer biologist who described with amazing accuracy what only now, three decades later, are finally accepted as real and identified as *transposons* by mainstream molecular biologists. When Evelyn Fox Keller asked her how it was she could look through a microscope and see in organisms what others could not, McClintock replied, "Well, you know, when I look at a cell, I get down in that cell and look around."[195]

Every day we use expressions such as "a different point of view," "a second opinion," "another perspective" — expressions which reveal a commonplace understanding that the point of view, the perspective, of a single person is often insufficient to give us a good idea of what we are actually seeing, reading, or hearing. We all know that to understand something fully, or to see something completely, we need to look at all sides of it. An architect, for example, provides her client with different perspectives of the building being designed, and now computer design can provide every perspective, not just three, by rolling around the virtual building to whatever point of view is desired.

In other words, we all recognize that there is more than one "way of knowing" *(epistemology)*. Each of us sees and experiences the world in a slightly different way; of necessity from a different, and unique, standpoint, not only because we each occupy a different space, but also because we are each unique organisms. We cannot quite see or experience exactly the same thing, so when we compare experiences, our descriptions will never be identical.

But an epistemology is more than a perspective. It is a system of knowledge.

I first encountered the word "epistemology" in a theological context, referring to ways of knowing God, when I began to emerge from the isolation of our farm and started to read the feminist critique of both science and theology. It took me a long time to grasp the significance of the word and become comfortable with it. After all, I grew up in the US where science is science and truth is truth. There is only one, and we have it.

Epistemology, says Steven Rose, "refers to how we study and understand the world," while *ontology* "refers to our beliefs about how the world 'really' is."[196] This is a useful distinction, since it suggests that we can believe the world is a unity, even a "sensible" place, and can thus dare to act and to live in it. Rose describes this material world as "an ontological unity ... which we approach with epistemological diversity." Imagine, for example, a volleyball game. Before the game starts, we agree that there is one ball, one net, and one set of boundary lines (ontological unity), while each of the players sees and experiences this in unique ways (epistemological diversity). "We require epistemological diversity in order to understand the ontological unity of our world," says Rose. We also have to recognize that there are others who will not recognize, or be able to "see," our nets, boundaries, and rules.

Western science may be able to provide a chemical analysis of a plant leaf but be unable to tell us anything about its healing properties. A tribal person in India or a Costa Rican aboriginal may not be able to speak English, may have no training in reductionist science, and may have no explanation (or even interest in having an explanation) as to the "what" of a plant in terms that we can recognize, yet may know not only the healing properties of the leaf, but also the nutritional qualities of the roots and nuts as well as its growth habits and sensitivities. The culture of the west, however, is historically a monoculture, and in its traditional form there is little recognition of, or place for, differing epistemologies or cultures. It is the hubris (overweening pride) of western civilization to claim superiority and exclusivity for its unique brand and package of knowledge, including its science. It feels no need to entertain or even acknowledge other truths, understandings, or insights. Industrial biotechnology and genetic engineering are artefacts of this culture.

One aspect of epistemology is *what* we know. Another is *how* we know it. Do we come to know the properties of a plant or a molecule or a person by taking them out of context and isolating them so that we can reduce the variables before we dissect them? This is the fundamentalist approach of reductionist laboratory science. But what we can know is thus severely limited and perhaps not even reliable or accurate because no object, no organism, is without context or environment. The genetic code, or the slice of DNA, without context could mean many things, or nothing.

When a child cries, we could choose a reductionist approach to determining the cause and take a blood sample for analysis, submit the child to a brain scan, or administer a psychological test. Or we might choose to look at the child's context and observe that there is a monster peering through the window or that the child ate too many cherries or that she has a finger jammed in a door. We could also extend the idea of context in time, to consider what happened to the child in infancy, the previous day, or what he is facing in an hour when taken to the hospital.

How is it then that we have attributed the divine attribute of total comprehension — omniscience — to what has become a monocular reductionist science? Reductionist western science is just one way of seeing the world, a single perspective, however it dresses itself up.

If we turn the question of perspective inside out — if we look out instead of in, or at, an object — we find ourselves looking at, or considering, context.

We all grow and live within a particular context. Every organism and every cell of an organism has a context without which it does not exist. This context can best be thought of as a living system which is constantly changing, constituted by "living organisms and their trajectories through time and space," trajectories that Steven Rose refers to as "lifelines."

Another way of situating an organism is described by philosopher and teacher Craig Holdrege. A plant is never whole in space alone, he says: "The whole is created in time .... The 'time-body' of the plant engenders its 'spatial body'."[197] The lifelines of an organism reach laterally in space and forward and back in time, situating the organism, establishing its context.

From where I now sit I can look out and situate myself spatially. I can also recall my past and imagine the trajectory of my life into the future, thus establishing my lifelines. To put it more personally, I can recall the people around the world who contribute to the definition of my life, my organic self, in the present, and I can also remember the people who have, in the past, given shape and direction to my life and who are, certainly for me, lifelines. There are genetic lifelines and social lifelines that include my parents, my children, various aunts and uncles, teachers, sages, prophets, friends — and some I would never regard even as friends.

Holdrege's statement that "Contextual thinking knows no isolated objects, the 'things' of the world recede and relationships reveal themselves," reminds me of my distaste for grade school history as a reified ("thingified") chronicle of kings and queens, wars and victories, treaties and inventions. When I finally experienced the teaching of history, halfway through university, as a relational account, a story, of cultures and conflicts, classes and consequences, a new old world opened to me. History, now filled with lifelines, came alive to me as an organismic whole, more as myth than chronicle. The difference was that between understanding persons by dissecting their bodies and understanding persons by studying their social relations. Biotechnology is of the former school.

In the course of analyzing the history of social attitudes toward science, Jon Turney describes how the biologist, through the first half of the twentieth century, was represented as a man in a white coat, armed with a microscope and glass dishes or test tubes. After the mid-1950s, however, "the white-coated figure" was most often posed with a molecular model or a blackboard diagram of the double helix. More recently the image has changed again. Now the object of study is "a stretch of printed DNA code, or a photostrip of a DNA fingerprint — *the emphasis on information rather than structure*"(emphasis in original).[198] In other words, the object of study, the organism, becomes increasingly isolated and devoid of context.

In this sense, the industry is right about there being nothing novel about biotechnology; already in the mid-nineteenth century "a new breed of physiologists asserted that life could be explained in purely physico-chemical terms...Vivisection was not new, as Victor Frankenstein testified when he told Walton how he 'tortured the living animal to animate lifeless clay'. But its large-scale use was. It was now an integral part of an interventionist, empiricist approach to biological problems."[199] Biotechnology has just taken reductionism and dissection further, into ever smaller pieces.

"Corporate and academic research in biology have become 'biotechnology'," writes Richard Strohman. "There are no organisms in either; there are no wholes but only mechanisms and parts, and the overriding concept guiding this research is 'fundamental' or 'basic' only in the futile hope that some day the parts and mechanisms will somehow add up to the whole."[200]

Strohman's comment on structure (or lack of "product") is complemented by Steven Rose's comment on the process of western science: "The successes of science have been based not so much on observation and contemplation, but on active intervention in the phenomena for which explanations were being sought ... we cannot escape the fact that interventionist biology, and above all physiology, is a science built on violence, on 'murdering to dissect'."[201]

Reductionist science and the culture of control, which form the foundation of industrial biotechnology, eliminate context of necessity. Picture the traditional Russian doll, starting with a large *babushka* that comes apart to reveal another *babushka* inside and so on down to a thimble-size solid wood *babushka* — the gene. Context after context has to be removed to discover the essential information of what is purported to be life. But what would we make of the solitary, timeless, solid wooden *babushka* if that was all we had? Maybe a better analogy is an onion: painfully strip away layer after layer and there's no onion left — nothing but sore eyes!

Industrial biotechnology, like the industrial form of agriculture to which it is being applied, only acknowledges context as a problem to be managed or as a restriction from which to be liberated. The talk about precision; the assurances that only a single gene is being altered, deleted, or added; the bluster of pronouncements that this or that cannot happen — all deliberately ignore lifelines or any contextual dimension either as influence or as consequence. Cargill, for example, began selling hybrid sunflower seeds in India with the proud claim that it used only the best germplasm from around the world — none of it Indian. And transgenic soybeans designed to be grown in Ohio or Indiana or Ontario are sold around the world, as if the seed neither has, needs, or recognizes any particular context.

But the organism as an automaton is a fiction. "Boundaries between organism and environment are not fixed. Organisms are constantly absorbing parts of their environment into themselves as food, and are constantly modifying their surroundings by working on them, by excreting waste products, or by modifying the world to suit their needs," says Rose. Organisms are not simply passive responders to their environments. They choose to change them and work to that end. "Organisms ... are active players in their own futures."[202] In other words, organism and context are always interacting and constantly changing. This is life, not the isolated DNA with its "information," not the molecule, not the cell of an organism.

Without context, death

Modern biotechnology is either an extreme expression of the nine-teenth-century vivisectionist fascination with the mechanisms of life that required the death of the subjects of study, whether stray cats or condemned criminals, or it is a radical departure in its ability to transgress boundaries and violate the integrity of both dead and living organisms, from bacteria to human beings. On the other hand, it might be both.

"Western science has made a better organized attack on the secrets of nature and used greater resources in the assault than science in other cultures," claimed the authors of an article in *Scientific American* a number of years ago, citing the "western economic miracle" as proof.[203] The claims may remain true, but one does have to wonder whether western science in general, and biotechnology in particular, will follow the markets downward and produce as much chaos and social destruction as has the market's descent since 1998.

Whatever the judgement of history, the public and scientific voices of dissent from the project of biotechnology have become more profound and more numerous as the industry claims of not only knowing the "secrets of life," but owning them as well have become more sweeping, strident, and offensive.

Biotechnology is hailed as the engine of the economy, the only hope for feeding the hungry of the world and the only way to save the environment from human desecration, but there is still another seductive promise for those who can afford it: the possibilities of defeating death, or at least putting it off.

The "scientific impulse" that drives biotechnology is described by Evelyn Fox Keller as "the urge to fathom the secrets of nature, and the collateral hope that, in fathoming the secrets of nature, we will fathom the ultimate secrets (and hence gain control) of our own mortality."

This "campaign," as Keller puts it, proceeds on two fronts: "The search for the wellspring of life, and simultaneously, for ever more effective instruments of death."[204] The success of this venture, and the recognition of molecular biology as a true science, are marked, she says, by the discovery of the structure of DNA on the one hand and the making of the atomic bomb on the other.

Keller presents a compelling argument that casts western science as a male project to overcome the secrets of life held by women. "Secrets function to articulate a boundary, ... a sphere of autonomous power. And if we ask, whose secret life has historically been, and from whom it has been secret, the answer is clear: Life has traditionally been seen as the secret of women, a secret from men, [b]y virtue of their ability to bear children." Western culture, says Keller, intentionally "invented a strategy" — the scientific method — to deal with "the threat or the allure presented by Nature's secrets" and to assert power over "Nature's potentially autonomous sphere."[205]

Not surprisingly, it is mostly women, such as Maria Mies, who seem to notice, and identify, the violence inherent in this attitude and practice. "Without violently disrupting the organic whole called Mother Nature, without separating the research objects by force from their symbiotic context and isolating them in the laboratory, without dissecting them ... into ever smaller bits and pieces, ... the new scientists cannot gain knowledge. They cannot, it seems, understand nature and natural phenomena if they leave them intact within their given environment."[206]

The search ever deeper into the organism for the fount of life has found a new focus in the embryonic stem (ES) cell, as it is referred to by those who hope that this ES cell might be the "factory in a dish" that could turn out spare human parts on demand. Of course, "this would also be a potential gold mine for the biotech firm that took out an enforceable patent on the *tabula rasa* cell."[207] Before biologists can study this primordial cell, however, "they need to capture it — and control its growth." To do this, they need to isolate it. In other words, they have to remove it from its context before they can study it. What then is it that they are studying if, as we have seen, an organism, or its constituents, can only be fully identified and understood in its life context?

Furthermore, Keller points out, "As the search of particle physicists for the building blocks of matter leads them into the realm of the vanishingly small and evanescent, the search of biologists for the building blocks of life leads them into the realm of pure information,"[208] information that seems to slide away just as soon as it is within grasp.

The result is, as Strohman comments, that what began "as a narrowly defined and proper theory and paradigm of the gene" gets pumped up and mistakenly evolves into a theory and a paradigm of life. "Normal science is an approach that reveals genetic maps related to biological function, but the directions for reading the maps are not included in the package. And the real secrets of life are obviously in those missing directions .... These rules are more than likely embedded in the organization of life rather than in the catalogue of the organization's agents."[209]

The biotech industry views it all in a rather different way. "Each of humankind's estimated 100,000 genes contains a precise molecular script for making a different protein .... It's the role of proteins that scientists need to know to unravel the secrets of life and to develop potent new drugs."[210]

The elusiveness of the alchemists' dream is reflected in the seemingly constant receding of the goal over the horizon. First the prize was the structure of DNA, the discovery by Watson and Crick that was actually pirated from the work of Rosalind Franklin. (The scenario would have been radically different if Franklin had followed contemporary practice and patented her discovery before anyone else knew about it. Francis Crick might have remained just a high-level lab technician.) Next it was the sequencing of DNA and the identification of specific genes. Then it was the identification of what specific sequences coded for. Now we are in the midst of a more substantial relocation of the dream into the proteins, but Ruth Hubbard reminds us that "proteins and genes have a sort of chicken and egg relationship. Many genes are implicated in the synthesis of any given protein, and many proteins are involved in the synthesis and functioning of any given gene."[211] With an estimated 100,000 proteins produced in the human body, there are 50 billion possible protein combinations.

One has to wonder where the secrets will be hidden next. One has also to wonder when reductionist science will recognize that its whole paradigm is wrong, as Strohman (among others) has suggested; that single objects, sequences, proteins, or anything else are not the secret of life, whatever one calls them or wherever they are found.

The search for the secrets of life faces a fork in the road: it can shift its focus to the social relations and lifelines of organisms, or it can continue to seek the essence of life in some vitalist fashion, believing that if an organism is sufficiently dissected, the secret of life — the "ultimate cell" — will at some point stand revealed, exposed to control.

Such a revelation, if possible, would be at the cost of sacrificing life itself.

⅋

With the paring and peeling away of life has been a paring away of death, whether from plagues or pestilence, with highly ambiguous consequences. Only a few years ago we were assured that malaria was on the verge of being eradicated, along with TB. Antibiotics would eliminate all sorts of opportunistic pathogens. It was to be only a matter of time before, one after another, the diseases and demons that inappropriately terminated human life would be eradicated — put to death by human ingenuity and modern science.

The next step would be to eliminate — or be able to repair — the ravages of those less-than-terminal ills that beset human kind. "Organs without donors — that's the logic behind fusing cow eggs with human cells .... It's a transplant surgeon's dream, an endless supply of organs and tissues neatly matched to their recipients .... Because these tissues and organs would be cloned from the patient's own cells there should be little problem with immune rejection. They could provide a desperately needed solution to the chronic shortage of donated human organs ...."[212]

Science is increasingly treating us as automobiles, consisting of globally sourced components accessed through its exclusive dealer network — which is exactly how the biotech industry already regards plants.

I recently received a new "gold" provincial health card, my ID as an over-65 "senior," along with a form for organ donations from the BC Transplant Society with an envelope printed with "postage generously paid by <Roche>" in the upper left-hand corner. I, too, could contribute to the just-in-time industrial manufacturing process developed by the Japanese automakers, in which components, rather than being produced and stored in a warehouse to be used when needed, are manufactured in sub-assembly lines (bodies like mine, in this case) anywhere in the world and shipped to arrive just as needed in the operating room. The sub-assembly lines might be plain folks like me — or they might be the street children of Brazil or the women of India impoverished by Progress. Soon pigs or other animals, specially engineered for the job, will be producing spare parts (xenotransplants) — and perhaps deadly viruses to keep them company. In any case, Roche will be a winner, even if we are not.

But is this any more than the latest chapter in the long and rather sordid history of our culture's attitudes towards children and our apparently age-old *Quest for Perfection?* In her book of this title, Gina Maranto describes the various ways unwanted — or imperfect — children were disposed of in the past. As for the present, she says, "Inarguably, control is the *sine qua non* of science .... The whole point of the assisted-reproduction enterprise, that culmination of those two thousand years of persistence, intuition, and inventiveness, is to exert mastery over procreation, to remove chance from a natural biological process. Already, reproductive endocrinologists and others in the infertility business are participating in eugenic decisions made by couples employing their services. One day, if they have their way, scientists-cum-physicians will replace faulty genes in embryos as if they were bad carburetors."[213]

The reproductive technology market may be even richer than the death-and-disability-avoidance market being created by the biotech industry. The fact that people unable to bear children feel — or are made to feel — inadequate, less than human, and unwanted, is used to legitimize activities designed to produce a child regardless of the condition of the parents or the people seeking to produce a child. The transformation of a desire, or a longing, into a "right" hands power to the baby-manufacturing division of the biotech industry. Again the problem is defined as individual and genetic, not social and environmental, for the benefit of business. An extreme expression was a suit brought by a Nova Scotia lawyer against the provincial government. He demanded that the health care system pay for expensive and seldom-successful high-tech *in vitro* fertilization (IVF) for himself and his wife as "a necessary medical treatment." "Infertile people are lying in ditches on the side of the road and they're bleeding because their hearts are broken," said Alex Cameron in his closing statement.[214] Early in 1999 the judge ruled that "this is not a medical end and in this matter the medical procedures used to attempt to have a child...have not been shown to me to be 'medically required'."[215]

The confusion of life and death is illustrated by another brief story: "The man died suddenly, and his family asked for his sperm to be preserved." A team of doctors squeezed the sperm out of the dead man's epididymis and froze them. The supervising doctor said, "It gives people hope and lessens the pain of suddenly losing a loved one." Later, after being defrosted, the sperm were injected into eggs harvested from the man's wife.[216]

The esteemed biologist E.O. Wilson has recently written, "To the extent that we depend on prosthetic devices to keep ourselves and the biosphere alive, we will render everything fragile. To the extent that we banish the rest of life, we will impoverish our own species for all time. And if we should surrender our genetic nature to machine-aided ratiocination, and our ethics and art and our very meaning to a habit of careless discursion in the name of progress, imagining ourselves godlike and absolved from our ancient heritage, we will become nothing."[217]

Chapter Eleven

Apocalypse, Armageddon

Ending and beginning

Smouldering below the surface of genetic-engineering biotechnology, and erupting in alarming ways at times, is a vein of metaphysics that provokes a number of questions about the real nature of the business. Whether it is the costly effort to develop replacement parts for human beings, develop techniques of genetic selection for the "perfect" baby, "improve" canola or corn for whatever supposed reason, or Monsanto's offering "food - health - hope," questions about the secrets of life and death are imminent.

Monsanto's screenplay for its own biological mutation and metaphysical transformation from an evil chemical company at the end of one millennium to an angel of life with the revelation of the savior for the new millennium, is blasphemous and pretentious; fundamentalist scenarios of Armageddon need to be.

On the other hand, some critics suggest, it may be that Monsanto believes none of its own propaganda (why should it?) and that it knows that it is just a matter of time before this company, or some other, produces an ecological if not human disaster. If such cynicism does indeed rule, then it is a simple matter of seeing what you can get away with, and how much money you can make, before the Apocalypse.

Millenarianism is the expectation that the end of the world is near

169

and a new earthly paradise at hand. This is often interpreted as a restoration of the Garden of Eden; historian David Noble, for example, in *The Religion of Technology*,[218] focuses on Joachim of Fiore's interpretation of the Book of Revelation and his claim (in the year 1200) to be able to read the signs of the times and thus predict events yet to come. "Armed with such foreknowledge, which included an anticipation of their own appointed role, the elect needed no longer to just passively await the millenium; they could now actively work to bring it about." Technology became a means of grace to achieve this.

Noble argues that a belief in an approaching millennium and "the now long-standing hope of recovering the Adamic knowledge lost with the Fall" was at the heart of seventeenth-century science. He puts Francis Bacon in this camp: "Bacon's advocacy of the useful arts in the interest of advancing human knowledge was aimed above all at the fulfilment of the millenarian promise of restored perfection." But, as Noble points out, "despite their devout acknowledgement of divine purpose in their work, the scientists subtly but steadily began to assume the mantle of creator in their own right." By the end of his life, Bacon was predicting (in *New Atlantis*) that "men would one day create new species and become as gods."

That was three centuries ago!

Noble goes on to discuss the "revived millenarianism" of the nuclear age, which was based upon a renewed belief in both inevitable technological destiny and deliverance, or "salvation," by way of an atomic Armageddon. It is hard, now, to call to mind the political and emotional climate of the Cold War era, the three decades from the 50s through the 70s. When the biotech promoters speak now of the hysteria of their opponents, they are speaking in ignorance of the hysteria that lurked just below the surface in millions of people of all ages during the days when a fire siren could freeze teenagers as they awaited the worst. I still recall that experience with a chill. It was the willingness of the demagogues — politicians and presidents, Kennedy among them — to destroy the world to "protect" some ideological construct called "freedom" (today it would be "the market") that was so frightening.

Still we got used to it, with the aid of the increasingly sophisticated rationalizations that were laboriously fabricated by theologians, politicians, and think tanks such as the Hudson Institute. Now the Hudson

Institute provides institutional legitimacy to Dennis Avery and his "pesticides and plastic" — and now biotechnological — utopian fantasies. "We" won the propaganda war, and the Russians are paying for that victory today.

The Cold War is history, and we have forgotten about the high-stakes game that was played, which included preparations for CBW — Chemical and Biological Warfare — just in case the nuclear deterrence trick did not work. CBW preparations were the direct antecedent of the current campaign of germ warfare, and I was active in the resistance to all forms of such madness.

Etched in my memory — and marking my soul to this day, I am sure — is the experience in 1960 or '61 of standing in silent vigil at the gateway to Fort Detrick, the home of the US Army's CBW research and weapons facility in Maryland. (It is now implicated in biotechnology pursuits such as the Human Genome Diversity Project.) The vigil was organized by a Quaker couple who maintained a silent vigil at the main gate to Fort Detrick for several years, in protest of the evil activities being engaged in there and as an appeal to the employees who passed by twice each day. Standing in silent witness, meditating on the potential of what was being prepared in that dreadful place, was an eerie and disturbing experience.

There were other vigils. While I was in seminary in New York there was the annual farce of a grand, one-day, civil defence drill, when everyone was expected to practise hiding from nuclear bombs. A small group of us from the seminary took the opportunity to conscientiously refuse to cooperate. We made a display of going out and sitting on the steps of the Lincoln Memorial in Riverside Park. Downtown, Dorothy Day and the Catholic Workers would get arrested in routine fashion. Uptown, the cops on patrol would pretend not to see us.

Now, perhaps, we should be thinking about the drift of transgenes from Roundup Ready canola in the same way we once thought about the drift of nuclear fallout. But how does one shelter the plants in the field?

Biotechnology has "progressed" from being a tool used to create new weapons of silent mass killing to being a technology that creates new products as instruments of salvation. No such claims, as I pointed out in Chapter Six, were made for the first retail products, the Flavr Savr tomato and recombinant bovine Growth Hormone. To compare them with nuclear bombs would have seemed absurd — at the time. But maybe they were just the opening salvos in the current campaign of germ warfare.

Monsanto CEO Robert Shapiro "does not forecast an apocalyptic collision between growth and the earth's 'natural limits'," reported the *Economist*. "Some say Mr. Shapiro's seeds are the biological equivalent of nuclear power — another product once acclaimed as a clean alternative to the depletion of the earth's resources."[219]

The millenarian vision of the restoration of the Garden of Eden, or at least of the paradise Adam was enjoying until he got too greedy, may have motivated nineteenth-century scientists, but with the secularization of western society, the myth has lost its power and a new mythology — a not-so-drastic revision of the Frankenstein story — has had to be created for the enterprise of biotechnology. This new myth, now on offer from the transnational "life sciences" corporations and expressed most succinctly by Monsanto's new logo, does not offer a return to Eden, but the millennial fulfilment of the serpent's promise that if Eve would eat of the Tree of Life, her eyes would be opened "and she would be like gods, knowing good from evil." In the new millennium, however, the corporate Eve has patented this apple/knowledge/information and now insists that you pay a royalty for a fragment of her knowledge of good and evil.

Frankenstein, Jon Turney points out, "marks a transition, in stories of men creating life, because Victor does not invoke the aid of the Deity, or any other supernatural agency. He achieves his goal by dint of his own (scientific) efforts."[220] These efforts, however, depended in part on the knowledge acquired through the dissection of human bodies, however obtained. So Mary Shelley has Victor say, "To examine the causes of life, we must first have recourse to death. I became acquainted with the science of anatomy: but this was not sufficient; I must also observe the natural decay and corruption of the human body." The end of the story, of course, is that Victor's success in the creation of life ends in tragedy and death.

&

In a succinct account of what he refers to as the specifically Christian character of western technology as it emerged during the Renaissance and Reformation, Ernst Benz points first to the Christian — and Old Testament — concept of God as Creator, specifically as potter and master builder, and then to the evolution of God into the great machinist or clockmaker, author of the cosmic machine, in the context of the Industrial Revolution.[221] In the words of Evelyn Fox Keller, "Nature, relieved of God's presence, had itself become transformed — newly available to inquiry precisely because it was newly defined as an object."[222]

Describing the world as created and God as Creator deprives the world of its divine character: it is not God, but made by God as something temporal and transitory. Humans are then created in the image of God, says Benz, according to one Genesis story and tradition, and placed in the world to care for it — or, as it came to be interpreted, have dominion over it. In the very act of creation, the world is alienated from God, it becomes not-God, and Adam is subsequently created as an alienated creature — alienated from God, separated. It takes only a slight twist of logic to argue that if humankind is made in the image of God, then humankind must also share in God's creativity and be co-workers with God in the establishment of God's kingdom. Finally, as the logic develops, God is quietly marginalized and disappears altogether. The completion of Creation becomes a uniquely human enterprise aided by modern technologies: the Frankenstein myth renewed.

And what better, or more powerful, technology to achieve the completion of life — or to bring about Armageddon — than genetic engineering.

Pursuing a different line of historical analysis than Noble, Benz suggests there is one more powerful element to be added into the story: the idea that God the Creator was also the Lord of a history of salvation that is working toward a final goal. This is the foundation for the idea of progress: history is subject to a divine plan that is progressing toward the completion of Creation, with humankind as the agent of its completion. "The modern technological revolution has never been able to deny its eschatological roots, its determination by a Christian understanding of time and history. It converted the Christian expectation of the coming of the Kingdom of God into a technological utopia."[223]

Benz's analysis leads us logically into the contemporary culture and dynamic of biotechnology, even though, like Ivan Illich, he wrote at least a decade before the dawn of the era of biotechnology (this book was first published in German in 1965), when he could not have foreseen the utopian project of perfecting life through genetic engineering. His reasoning, however, provides at least some insight into the cultural hubris of western reductionist science and industrial culture.

The final steps in the process Benz describes are provided by Bruno Latour, who suggests that in the process of modernization, God crossed from a position of immanence in nature to immanence in humanity. Nature was thoroughly secularized, isolated from God and any spiritual forces, while humanity became spiritualized, endowed with an uncritical inward presence of God that provided sanctification without critique. "He [God] would no longer interfere in any way with the development of the moderns, but He remained effective and helped within the spirit of humans alone."[224]

This reasoning provides a plausible explanation for the hubris of the modernist project of biotechnology, in which intervention in nature is unrestricted, and it underlies the understanding of science that it cannot be subject to critical or ethical considerations. One does not stand in God's way!

> For whoever finds me finds life, and obtains the favor of YAH-WEH; but whoever misses me harms himself, all who hate me are in love with death. Proverbs 8:35-36 (Psalm 8)

Word and seed

> "In the beginning was the word."
> But was it *ex situ* or *in situ*, written or oral?

Having looked at some of the practices of biotechnology and its attitudes and contrivances, and toured through the culture that has spawned this extreme expression of Western culture, there is now a fundamental question to be considered about the culture of biotechnology itself: is it dead or alive?

As I have noted, when Mary Shelley wrote *Frankenstein* in 1831, experimental biology was just getting underway. Dissection (of dead organisms) and then vivisection (of live animals and organisms) were becoming the mechanisms of choice for understanding the organism; this was the tradition of "learning about the body by taking it to pieces."

The public, particularly in Britain, was not comfortable with this violent assault on animals, and the anti-vivisectionist movement became a part of the social and political landscape.

The new "technology" and the attitude it expressed displaced phenomenological observation that had characterized biological research and teaching, including medicine, up to that time. Coupled with the mystical search for the secrets of life, the ground was prepared for the emergence of molecular biology a century later, followed by genetic engineering, as the tools became available to delve deeper into an organism, piece by piece, in search of life. (One does have to wonder what the practitioners thought they might find.)

The Frankenstein story, as Turney points out, is grounded not in the supernatural, but in the contemporary science of the time, which is depicted as the search for "control over the fundamental biological realities of life." As a horror story it is read and reread, but it has also taken on the character of myth as it expresses fundamental human hopes and fears and undergoes constant elaboration in the process of being told and retold, filmed and dramatized, thus remaining contemporary. "A myth is flexible, precisely because it is a characteristically oral form, but retains a stable core of meaning. A printed literary text, on the other hand, is fixed in form, and the critics' job is to discover its many possible meanings."[225] This is highly suggestive of the issues involved with genetic conservation, seed banks, and subsistence agriculture.

It was only a few years ago that concern about the conservation of seeds — "genetic resources" as they are now referred to by the biotech industry — focussed on *ex situ* conservation: seeds are removed from their native habitat and stored in modern refrigerated seed banks, usually in the affluent countries of the North. In other words, the "information" was put into cold storage, where it could be preserved as a dead artefact and some of it removed from time to time to be used as a novel source of genetic material (germplasm).

This "information" can also be thought of as a written text, something like the Dead Sea scrolls or some other ancient parchment. The presumption is that seeds are fixed and unchanging, and that the seed put into the seed bank five decades ago will have the same validity today as it did then. The seed, or more importantly, its genetic information, is treated as without context. The seed is regarded very much as a king or queen or the date of an important battle was in my grade-school history books: disconnected and dead. And of course what was recorded as history was always a highly selective account of events as seen by the victors. The losers are written out of history, just as other cultures, epistemologies, and cosmologies are. Scholars may spend many hours, even years, trying to ascertain the accuracy of historical accounts, but there may be serious inadequacies in the information they have to work with. It is very much like Dale Bauman extolling the wonders of rbGH and stating, "On the basis of the available information ...."

Contrast this with the practice of *in situ* seed conservation, which is the practice of conserving seeds in their native habitat where they remain in context and continue to evolve in a dynamic relationship with the world about them. They, and their keepers, may have to struggle against assaults such as the Green Revolution and genetic engineering, but they won't be wiped out by a power failure or budget cuts and negligence. The seeds conserved in this fashion thus have the character of myth. Their essential truths are maintained and conveyed by being planted, harvested, and replanted — told and retold as a form of oral history. There really is no space in such a culture for patent attorneys.

It has been said that "the vitality of myths lies precisely in their capacity for change, their adaptability and openness to new combinations of meaning."[226] One could just as well say "the vitality of seeds and living organisms lies precisely in their capacity for change, their adaptability and openness to new combinations of meaning."

Genetic engineering as a business, or an investment, requires that genes can be grasped, possessed, preserved, and exploited as commodities. "Would anyone think of investing in genetic engineering biotechnology if they knew how fluid and adaptable genes and genomes are?" asks Mae-Wan Ho.[227]

$\partial\!\!\!\delta$

A report on the situation in Bangladesh after the severe floods in the summer of 1998 underlines the crucial significance of the *in situ/ex situ* dichotomy. Farida Akhter of UBINIG, a peasant farmer organization in

Bangladesh, wrote that "when the flood water started receding at least from the homestead land, the farming families started making preparations for planting vegetables in their kitchen garden. Women were ready. They had tried to save their collection of seeds of pumpkin, beans, sweet gourd, lentils and various other crops. Some families lost the seeds, but women are ready to share and exchange among themselves. UBINIG took an initiative for seed distribution in the middle of September. Women are interested in receiving pumpkin seeds. All the pumpkin plants are gone. A house looks blank without a pumpkin plant. Rushia Begum, a woman farmer, was delighted to get pumpkin seeds. She expressed her happiness by singing a song, 'Praner Bondhu tomar Dekha Pailam' O dear friend, I am happy to see you." She was very happy because she did not have to buy from the market, which is usually only HYV or hybrid."

What complex understandings lie beneath such apparent simplicity, in contrast to the simple-mindedness of hybrid monoculture and the pretense of precision in biotechnology?

Akhter continued, "The whole selection process of seed is profoundly strengthening the indigenous knowledge of the community to cope with disasters. The selection itself is a highly technical task, and impossible for the formal sector to understand and rationalize because of the subtlety of the practice and logic to meet the diverse needs of the different households. So there is no one single variety, or one kind of vegetable, for every one. Each and every farming household has different and diverse needs. These are met through reciprocal exchange and appreciation of each other's specific needs. In case one family has a particular variety she can not plant now because of standing water in her field, she is offering the seeds to the other family."

In western culture, with its materialism and individualism, and its consequent emphasis on exclusivity, the written language is used to draw legal boundaries and define claims based on the notion of private ownership of property in the form of land, "natural resources," and now the "information" of life forms and processes. The seeds are patented and cannot be shared according to law.

Recently I heard it said that genetically engineered crops are creating a very quiet, very sterile agriculture; the crops do not share the land with weeds, insects, even bacteria. No sharing with anything. Since the farmer no longer owns the seeds he or she plants, they cannot be shared. The Terminator Technology — the production of patented sterile seeds owned by a corporation — is an extreme expression of this selfishness and greed. Will nature tolerate this affront?

Akhter describes how the sharing of seeds is part of the Bangladeshi farmer's culture. "They believe if you share seeds with your neighbor and friends, the yield will be higher. If the farmers keep seeds in their store, while other farmers have scarcity, then it will bring misfortune to the farmers since she/he deprives others. The culture of sharing indeed ensures diversity, it is a system that also ensures *in situ* conservation."

In his beautiful book *Enduring Seeds*, Gary Nabhan tells of the work of ethnobotanist Janis Alcorn, who described how "traditional farmers follow 'unwritten scripts,' learned by hand and mouth from their elders, that keep agricultural practices fairly consistent from generation to generation. Most land-based cultures have such scripts that guide plant selection and seed-saving. Each individual farmer might edit this script to fit his or her peculiar farming conditions, but the general scheme is passed on to the farmer's descendants. Thus the crop traits emerging through natural selection in a given locality are retained or elaborated by recurrent natural selection."[228]

At the heart of such stories and scripts is a devotion to the flow of life, not a calculating arrangement of information in a technological machine. Genetic engineering in its very conception requires the isolation and identification of fragments of life for purposes of appropriation, manipulation, and ownership. It must exclude the relationships and lifelines that constitute living organisms, ecologies, and societies.

The dominant and official culture of market capitalism imposes on all of us a choice between folkways and corpways, an essentially oral narrative that is by nature social and ecological, or a codified, written, legal construct designed as a contract of control in a centralized command economy. The commands, of course, emanate from the global headquarters of the "life sciences" corporations and are embedded in the genetics of the seeds, sperm, or eggs themselves, which, like colonies, are thus compelled to devote themselves to the needs of the "mother corporation."

Chapter Twelve

Growing resistance

It is impossible for the colonial situation to last because it is impossible to arrange it properly. — Albert Memmi[229]

Resistance: An underground organization struggling for national liberation in a country under military or totalitarian occupation; the capacity of an organism or tissue to withstand the effects of a harmful environmental agent. — *Nelson Canadian Dictionary* (1997)

I do not like to use military metaphors or language and I avoid the use of words such as "fight" and "struggle" unless their meaning is very precise. The notion of "totalitarian occupation," however, is broader than a military term and seems to be a reasonable way to describe the campaign of the biotech industry as it works to create a command economy of life.

If five or six giant corporations have control over every seed of all major commercial crops planted anywhere on the earth, that is totalitarian. Add to seeds control over the genetics of all major lines of commercial animals and it will be somewhat more totalitarian. Then engineer all the genetics — plant and animal — to be hybrids, sterile, or both, and the achievement will be without question totalitarian. It will amount to the occupation of the land — the earth itself — by foreign troops and their local mercenaries. At the other end of the food chain there is a growing occupation of the land by a handful of global super-market chains, and an occupation of the supermarkets themselves by transgenic foods and food products, unlabelled, so that the public can-not identify the invaders and thus avoid and reject them.

A good way to check for any possible paranoia is to ask, at various times of the day and at different locations in the system, a simple ques-tion: For whose benefit? If the question elicits the same answer regard-less of when or where it is asked, one can reasonably suspect a totali-tarian situation. If the answer each time is "corporate control and prof-it," then you will know resistance is appropriate and necessary.

We should not be fooled into believing that the intent of engineer-ing the seed and occupying the land is to feed the world or save the environment; it is to gain control and create dependency. Like all impe-rial and colonial endeavors, its purpose is to gain the ability to exploit the resources of the colonized area and people for the benefit of the imperial powers. The new twist is that the imperial powers are now cor-porations, not states. (Actually, there are precedents: the enterprise of Christopher Columbus and the Hudson's Bay Company among them.)

In the name of Progress, these new powers would like us to believe that there is no alternative to their biotechnological project. They are simply the agents of destiny. We should adjust to their rule with grati-tude for their leadership and their efforts on our behalf, whether we asked for them or not.

They would also like us to accept their confusion about life and death. Genetic engineering is about the "improvement" of life through its reconstruction, but it is only the data that can be reconstructed, processed, and delivered, not life. A corporation cannot control life. It can threaten, it can intimidate, it can take you to court, and ultimately it can kill. Like the state, it may have the power to take life, but neither have the power to give life.

For millennia, conventional farming has recognized the complexi-ties and diversity of being alive. Until very recently, the practice of agri-culture has been about nurturing those complexities, not about sim-plification and the eradication of diversity in the name of protecting a shrinking elite of genes, seeds, and people. The protection of life has been based on the sharing of seeds, plants, animals, land, and water, not the exclusion of others from what some have appropriated as their private property.

Today the transnational corporations are spending hundreds of millions, if not billions, of dollars to overcome any hesitation the public in general and farmers in particular might have about embracing the new technologies of domination and profit. Proprietary genes, germplasm, seeds, and embryos are the cluster bomblets in the war against life.

> In addition to more than two million tonnes of bombs dropped by the Americans [during the years of warfare three decades ago against the people of Vietnam and their neighbors] ... there were also ground battles .... As a result, Laos has the most severe UXO [unexploded ordinance] contamination of any country in the world .... The most common killers are US anti-personnel cluster bomblets, which the Lao people call "bombies." The most frequently encountered ... contains 100g. of high-explosive and 300 ball-bearings embedded in its steel casing. Around 90 million of this type were dropped and up to a third remain unexploded.[230]

Will the story be similar, only both more subtle and more prolonged, with transgenic crops in farmers' fields? The bomblets maim and kill, but at least they do not reproduce.

Biotechnology is not the science of life. It is a technology of violent intervention, domination, and death. It is an artefact of a culture that finds greater fascination in staving off death — or administering it — than in being alive.

&

During the time of growing student resistance to the war against Vietnam in the late 1960s, I saw the film *The Battle of Algiers* while attending a student conference in Ohio. It is the story of two men who grew up together in Algeria and were lifelong friends, even though one was a French colonial and the other an Algerian. The film centres on the situation they faced in the midst of the Algerian war for liberation from French colonialism in the 1950s.

The Algerian and the Frenchman come to the realization that history and circumstance have put them on opposites sides in war. If it becomes necessary, they recognize, they will have to face each other as enemies, prepared to kill each other. There is no longer any middle ground. In recognizing the tragedy of their situation, their respect for each other and their friendship is intense.

I remember the film vividly, and still with pain, because just after that I came to a similar point of no return with a man who had been my friend from our days together in kindergarten. I have never seen him since then because I realized, while visiting him the day after seeing the film, that it was the same for us. He had become the enemy, by choice of business activity and social standing or simply by class, and I realized that if there was a war, we would be on opposite sides. Yet we shared so much, in culture, in history. This is not a line to be drawn lightly.

There is, perhaps, an instinct in all of us — and certainly among Canadians — to seek the middle ground and avoid confrontation. I know that I do not like confrontation. But perhaps we extol the middle way simply because it seems like a useful way to avoid coming to terms with uncomfortable reality.

With biotechnology there is no longer any middle ground. It is a matter of life or death as the directors and their corporations intentionally polarize the issues and the discussion, as we have seen, with their fundamentalism. If you do not accept biotech, you are against it. Biotechnology has been turned, by its promoters, from a science into a cult. You are a true believer, or you are an infidel.

It is a sad situation, particularly for all those people who love the science — and may have spent years working in it, whether in corporate labs, university graduate schools, or government regulatory agencies and research facilities — but can no longer tolerate what it is being used for and how it is being manipulated. Elisabeth Abergel of York University, for example, tells how one day, while working as a molecular biologist for a biotech company, she was told to grow an unknown microorganism in order to characterize its genetic material. She asked where it came from, what it did in nature and if it should be identified first in order to know how to grow it, but was told that it didn't matter; they just wanted to know if there were any useful bits. That convinced her she was in the wrong place and should seek another line of work.

Such people deserve our support and assistance in finding ways to take their science out of the labs and into the fields where it can be put to the service of farmers and the public. It is like the underground railway that helped black slaves escape their American owners and, a century later, helped young Americans escape their country's war against the people of Vietnam.

Farmers and rural communities can replace the corporate boards of directors and their expensive lab-led science with farmer-led research using the resources of the community. The new "directors" will include plants, animals, and microorganisms as we learn to give them a place in the boardroom and to hear what they are saying to us.

Just as organisms or tissues develop the ability to withstand the effects of a harmful environmental agent, so do people. From affluent consumers in Europe to subsistence farmers in Bangladesh, the evangelists of biotechnology are meeting increasing resistance to their products. Even some regulatory agencies are losing confidence as they are forced to deal with the flaws and surprises that cannot be hidden, such as Bt potatoes that seem to have lost a major commercial gene, or confused canola that contains a wrong gene or two.

The resistance in the US does not have the support of the Rockefeller Foundation or the National Institutes of Health, and in Canada it does not have support from the Matching Investment Initiative, the federal tax credits, or any of the other monetary inducements with which the Canadian government encourages and supports the private biotech industry; but everywhere it does have the support of the quiet people who organize meetings, print and distribute leaflets, and feed and house the messengers and scouts of the resistance.

Steve Emmott of the European Greens has said that "Industry does not understand what we are doing because we are not getting paid for it. They do not understand doing something because it is right. Industry is nervous because they cannot control us."[231]

Resistance can spread among people as it does among weeds and crops. It is now common knowledge that weeds can and will develop resistance to herbicides, just as bacteria develop resistance to antibiotics. It is only a matter of time. It is also now common knowledge that the genes conferring herbicide resistance can quickly migrate from the genetically engineered crops into weedy relatives and who knows where else. Herbicide-resistant canola is appearing in places where it should not be, according to the companies producing the herbicide-resistant varieties. And while the regulatory agency (CFIA in this instance) remains silent, company representatives can be heard to say, "We always expected a level of natural outcross would occur within the species."

$\mathcal{E}^{\text{a}}$

Does the canola compelled to contain herbicide-tolerant "genetics" get angry at the violation of its integrity? Is its promiscuous tendency to spread its genes around, including those for resistance, its way of rebelling? Do the potatoes forced to replicate alien Bt genes object to their forced labour? Do the cows injected with rbGH object to the distortion of their metabolism?

We bear a responsibility to engage in resistance on behalf of all the organisms that do not have the means to resist initially. Although the Holstein cow may be a high producer, there are limits she would probably not choose to exceed. But she is not given the choice at the hands of dairy farmers who respond to Monsanto's beguilements, so we have to make the choice on her behalf.

The more aggressive the biotech apostles and their strategies are, the more they fear the growing resistance of organisms of all shapes and sizes. They used to talk about the public's fear of new technology, but it is now obvious that it is the industry that fears the public — so much so, in fact, that it has come to regard the public — people not under its control, by definition — as pathogenic or toxic to the practice of biotechnology.

The industry, as a result, seeks protection from the public. Mimicking the agriculture it has created that requires protection from its environment, the agrotoxin-biotech corporations now find it necessary to call on the courts and the police to protect them. When farmers in India pulled down the seeds plant that Cargill was building in Bellary, Karnataka state, north of Bangalore in 1993, Cargill called on the police to protect its facilities. Monsanto bought Cargill's global (except North American) seeds business in 1998, and before the end of the year it was demanding police protection in the face of growing farmer militancy against the corporation not only in Karnataka but in several other Indian states. A Bangalore newspaper reported, "The American Ambassador in Delhi has written to the Government of the State of Karnataka asking it to provide strong police protection to all American companies in the city. Pointing out the previous repeated attacks on American companies by 'miscreants,' the Ambassador has requested the State Government to create an atmosphere without fear and anxiety for them to work. He has stressed the need for a special security for companies related to science and technology ...."[232]

The industry also uses secrecy (the government calls it business confidentiality or proprietary information), denial of labelling, and massive propaganda. Its fear of the public has led the biotech industry to try to keep the locations of its seed trials secret. It has learned from experience that when the public knows where its activities are located, it faces overt forms of hostility.

Civil responsibility

During much of 1998, the genetic-engineering resistance in the UK used electronic communications as skilfully as the peasants of Chiapas to keep the world up-to-date on their actions. Here's an edited sampling of its reports:

> There are currently around a hundred test sites where 13 biotech companies are going through the motions prior to colonising European soil. So far around 36 have been set upon with broomsticks, sickles and weeding gloves.

> ———

> In Totnes, Devon, one field in particular was visited by 600 local people, some of whom closed their shops to make the trip. The field of GE maize, within pollination distance of the country's largest organic farm, was destroyed by approximately twenty people on a recent summer evening. Two of twenty have been charged with criminal damage and are subject to outrageous bail conditions. Local support has been amazing with over 300 people turning up at their latest court appearance and a petition with over 2,000 signatures from the local town alone. People signing are supporting the "illegal action" as it was in the public interest!

> ———

At 5 am, Saturday 23rd May, Britain's first Crop Squat began as 30 odd (very odd!) activists opposing genetic engineering moved onto a Novartis release site for experimental sugar beet at Kirby Bedon, near Norwich. Police arrived within 40 minutes and apologised for their initial hastiness; "Sorry, we heard there were 30 people in a field with sticks, but when we got here, we saw you were building wigwams!"

Within hours, organically grown flowers and vegetable plants had transformed this field of industrial corporate agriculture into an educational garden showing possible intercropping ideas and companion planting. The essentials of camp life, toilets, kitchen and camp fire were dug, built and lit respectively, and a well resourced information space was created in the yurt. Then the press arrived and carried on doing so for the next two weeks. Most of the support came from local people (some camping at night and going off to work in the morning). With an eviction order passed after two weeks in occupation (what do you expect if you squat the land of the Lord Lieutenant of Norfolk?) we packed up and left.

When the growing season and its accompanying festivities were over, the scene shifted to the supermarkets as they tried to cope with the mounting public pressure for a ban on all GE foods. As a growing number of major chains, led by Iceland [Frozen Foods], moved to rid their stores of GE foods, in early February 1999 the manager of the farming estate of Sir Timothy Coleman, Lord Lieutenant of Norfolk, announced the halt of all GE trials on the estate and ended its relationship with Novartis Seeds and Monsanto. The farm, as noted above, had been the site of Britain's first protest squat against GM crop trials, in this case, GE sugar beets. Estate manager Roly Beazley said, "There is a huge public debate over GE crops and a lot of public disquiet. Crown Point Farms does not wish to be in the middle of all this."[233]

The strategy of resistance, then, becomes one of isolating and disarming the occupation forces and decontaminating the occupied territory, a tactic highly developed in the UK. This is no easy task, but the resistance has the advantage of being decentralized, democratic, and motivated by more than an addiction to control and wealth. The forces of occupation, being aliens or mercenaries, are handicapped by their dependence on a highly centralized command structure and long supply lines. In other words, they are, appearances notwithstanding, vulnerable, which may account for their irrationality and their fear of the people.

Unlike the centralized and authoritarian structure of the biotech industry, the developing global resistance to genetic engineering is highly decentralized and spontaneous. There are prominent analysts and an elite of experienced activists working at the global level — Pat Roy Mooney and the Rural Advancement Foundation International (RAFI) may be the most well-known of these — but in a sense they are the research and support staff for the grassroots activists; they do invaluable work in obstructing and exposing the imperialistic projects of the biotech companies at the global level while the locals offer on-the-ground challenges and begin to build a different society.

For example, genetiX snowball in the UK, which arose during the summer in 1998, describes itself as "a campaign of nonviolent civil responsibility." Its position statement, broadcast by e-mail, expresses well the culture of the emerging democratic and non-violent resistance:

> After a century of facing the dangers of the nuclear age and the splitting of the atom, now we find ourselves near the turn of the century with the new, possibly even greater, threat posed by the splitting of the gene. Radioactivity has a "half life," meaning that it gradually becomes safer over thousands of years; but genetic engineering has "multiple life" — it keeps on replicating and can never be recalled once it is unleashed...
>
> genetiX snowball is a campaign of nonviolent civil responsibility which aims to build active resistance to the threat of a new gene technology which is unwanted, unnecessary, unsafe and irreversible. Our democratic system is failing us; transnational corporations have become too powerful and market forces are holding the reins of power. To put it simply, this means that profits are being prioritised over the health of people and environment.

The vision of the resistance is aptly expressed in this notice:

> The genetiX harvest action on Sunday 20th September 1998 has been changed from one of pulling up genetically engi-

neered (GE) crops to a "transformation" action. This will involve the planting of fruiting trees, organic seeds and vegetables as a vision of sustainable agriculture. The location of the transformation action, which will either be at a GE crop release site on a farm, in the grounds of a biotechnology company or at a GE research institute, will be disclosed at a press briefing ....

In natural biological fashion, the resistance is spreading:

> November 26, 1998: A group of conservative gardeners calling themselves the California Croppers held a tackle football match early Thanksgiving morning at the "Gill Tract" gardens in Berkeley, and in the process destroyed a crop of genetically engineered corn "as an informal welcome wagon gesture." The land is owned and operated by the University of California. The Croppers took the opportunity to welcome biotech giant Novartis, which had just signed a multimillion dollar research deal with UC Berkeley. The match was also meant as a Thanksgiving gift to Americans, who will be eating hearty meals today, 60-70 percent of which is estimated to contain genetically engineered food products. The rather pleasant and sporting activity of decontaminating fields of mutant crops is widespread in Europe and India ....

Given its Gandhian tradition and the sheer weight of population, with something like 600- to 700-million subsistence farmers, India is a logical place for GE resistance to take root and grow. In late November 1998, activists from the Karnataka Rajya Raita Sangha (KRRS — Karnataka State Farmers Association) announced that Monsanto's field trials in Karnataka would be reduced to ashes. Two days previously the minister of agriculture of Karnataka had been forced by journalists at a press conference to disclose the three sites where field trials with Bt cotton were being conducted in the state. KRRS contacted the owners of the fields and explained to them what action was to be taken, and for what reasons, and to let them know that the KRRS would cover any loses they would suffer. They also notified the world:

> Sindhanoor, India, 28 November 1998 — The direct action campaign of Indian farmers, Operation "Cremate Monsanto," started today in the village of Maladagudda, about 400 km

north of Bangalore. Mr. Basanna, owner of the field where an illegal genetic experiment was being conducted without his knowledge, and Prof. Nanjundaswamy, president of KRRS (a Gandhian movement of 10 million farmers in the Southern Indian state of Karnataka), uprooted together the first plant of genetically modified cotton, inviting the rest of the local peasants to do the same. Within a few minutes, all the plants in the field were piled up and ready to be set on fire ....

We are calling ONLY for nonviolent direct actions. Nonviolence in this context means that we should respect all (nongenetically modified) living beings, including policemen and the people who work for these TNCs.

The biotech industry is using intimidation, legal action, and threats of lawsuits, whether against farmers for infringing on corporate "rights" by saving seed, or against journalists and the public for talking openly about corporate affairs in a critical manner — or even for trying to find out what it is that made them sick. The corporate mind does not seem to understand that resistance arises from more profound concerns and convictions than what is to be reported in the next quarterly statement to shareholders.

Conscience

The first and most obvious step in resistance is always at the level of personal conscience.

When politics is subservient to the corporately controlled market, democracy may best be expressed by decontaminating fields and gardens or by action taken in the retail market itself. For the public, this means making conscientious choices about what is purchased and consumed, as well as where it is purchased or whether it is bartered or even grown at home.

The same holds true for farmers and gardeners and all those who want to take greater responsibility for feeding themselves, their families, and their communities. The first step is to shun all GE seeds and everything associated with them. The next step is to go back to selecting and saving your own seed and trading with others. This ancient practice is being revived more and more frequently around the world as people experience the biotech companies' tightening grip on the seed trade.

The retailers, wholesalers, and processors who rely on your dollar are very sensitive to public opinion, and your "vote" will be taken seriously. It can be cast not only by boycotting GE products or suspected GE products, but also by contacting the companies producing them through the addresses on the labels of the canola oil bottles, the bags of potatoes, the milk cartons, and all the corn and soy products. Copies of your letters, perhaps with an additional note, could even be sent to your elected representatives.

There is also the simple and civil action of talking directly with local store clerks and managers and translating personal choices into public actions, such as saying to the checkout clerk in a voice audible to those near you that you would like to buy this canola oil (or whatever), but not if it is genetically engineered or comes from GE crops. Unfortunately, you add, the product is inadequately labelled and does not provide this essential information. Such simple resistance can be undertaken with levity and respect and will generate some interesting exchanges as the poorly paid and uninformed cashiers and department managers realize they cannot answer your questions. I had a great conversation with a store manager one day about rbGH milk; he grew up on a dairy farm and his dad wanted nothing to do with the drug. On another occasion, after an exchange of letters, I had a lengthy phone conversation with the young woman in Toronto at the headquarters of the Weston empire (owner of Loblaws and Superstore) who called to ask me just what it was I wanted to know about their canola oil. I had to give her a basic introduction to genetic engineering so she could go to her bosses and find an answer to my questions. I am still awaiting her reply while the company decides what it is she should tell me. In the meantime, as the company gets more and more letters, their stores might have a visitation such as described in this British example:

> On 20th June, mutant vegetables — such as vegetables crossed with chickens — were on the loose in Safeways, Taunton, Somerset. Touring the aisles in a shopping trolley they declared their plans for global takeover, explaining to amused shoppers, "there's lots more of us you know, we're just the ones that got out first!" Once removed from the store the vegetables continued their warnings and gave out leaflets

which read "Thank you for taking part in our experiment." Over zealous staff then tried to bring the Veggies back into the store until the police arrived, but on hearing that it would be false imprisonment became aggressive until the vegetables decided they would call it a day.

k

When weeds develop resistance to herbicides, they do so from within for reasons of survival. You can almost hear the weeds organizing and saying to one another, "We are more entitled to be here than Monsanto's transgenic agents."

Resistance must begin within, as a matter of conscience. It is a matter of life and death, stark as that may sound. It is an issue, in the words of the prophet Moses, of life and death. "I am offering you life or death, blessing or curse. Choose life, then, so that you and your descendants may live ...." (Deut. 30:19) The prophets of biotechnology may offer life, so that you, personally, may live forever — if the companies just get the investment and the regulatory approvals to develop their technology. But the cost, and the price, is death.

Now is the time to replace the centralized command economy of the corporate world — which seeks to embed itself in the most fundamental structures of life — with a variety of decentralized democratic economies.

The time is overdue to replace the ethic of competition with the ethic of sharing, on the understanding that there really is enough to go around.

Appendix 1:

The Farmageddon Lexicon

Lexicon: 1) A handbook of definitions;
2) A process of putting tongue in cheek.

Armageddon: 1) The scene of a final battle between the forces of good and evil; 2) A decisive or catastrophic conflict. *(Nelson Canadian Dictionary)*

Apocalypse: 1) Great or total devastation; 2) A prophetic disclosure, a revelation. *(Nelson Canadian Dictionary)*

Attention Deficit Syndrome: A disease of the biotech companies, which can only see as far as the next quarter. Results in bad science.

Bio-Ethics: The limitation of ethics in matters of biology to the relationship between individuals, as between doctor and patient; the absence of social responsibility.

Biotechnology: The business of creating new products from living organisms. (Geoffrey Rowan, *Globe and Mail*, 1/5/90)

Burden of Proof: 1) An old moral concept being re-engineered through gene reversal; 2) You are guilty until you prove yourself innocent.

Concern: An artefact of the critics' minds. (Letter to the editor, *Nature Biotechnology*, July 1998)

Dangers: More real than risks; carefully omitted by the biotech industry and regulators in any discussion of genetic engineering.

Dependency: 1) A condition fostered by large corporations and imperial powers; 2) Being unable to sustain oneself without external support, such as food; 3) The creation of which is one of the purposes of genetic engineering.

Determinism, genetic: 1) "It's all in the genes;" 2) The genetic "information" we are born with; 3) We can only do what our genes are programmed to do, therefore we are not responsible for what we do.

Determinism, technological: 1) Technology is immaculately conceived; 2) Technology is "coming down the road"; 3) We have no choice but to use it, for better or worse; 4) Ethical and social issues are out of the question.

Engineered: Designed and constructed according to a preconceived blueprint or grand design; the product is designed before it is constructed, as illustrated in subdivision development, where all natural characteristics are eliminated so that the most profitable design can be imposed without compromise.

Eradicate: 1)   ; 2) 0

Familiarity: A concept adapted by the regulators of biotechnology to avoid offending corporate sponsors; if a food smells and looks like something familiar, then it is, and no further testing or analysis is required and it can go directly to market.

Farmageddon: Late-twentieth-century conflict apparently over control of crops and food, with prospects of turning into the final struggle between the forces of life and the forces of death early in the twenty-first century.

Gene: 1) "There are ... no genes .... A gene is a task that a cell has to accomplish" (Ernst Peter Fischer, cited by Craig Holdrege, *Genetics & the Manipulation of Life*); 2) "Each gene specifies the amino acid sequence of one protein. Indeed, that is what defines a gene." (Ruth Hubbard & Elijah Wald, *Exploding the Gene Myth*)

Genetics: What life has been reduced to, formerly known as seeds in the case of plants and sperm and eggs in mammals, including humans.

Germplasm: The mystical foundation of life, subject to private ownership under the rules of capitalism.

Genetically engineered organism (GEO): A word not used in polite company, particularly by propaganda machines such as Burson-Marsteller. See GMO.

Genetically modified organism (GMO): Euphemism for GEO. The Public is less apt to be disturbed by the term "modified" than by "engineered."

Gratitude: A healthy attitude toward life.

Hunger: A condition to be treated by genetic engineering if it can be done profitably enough.

Integrity: Inherent in all creatures, but missing from artificial life forms such as transnational corporations engaged in biotechnology.

Junk Science: What your antagonist says you have.

Junk DNA: The parts of a genome we do not understand.

Knowledge: Not the same as information.

Life: "The process of being alive." (Mae-Wan Ho)

Life Sciences: The assumed name of the campaign to reduce life to a prescription drug.

Modified: The violent transformation of an organism.

Monoculture: "The cancer of uniformity." (GRAIN)

Novel: Genetically engineered.

Novel Foods: 1) Would not be recognized as edible by Great-Aunt Sarah; 2) Foods that are not good for us.

Organism: A life form, regardless of complexity, with integrity.

Perceptions: Condescending description of attitudes adverse to biotechnology. Public concerns, even scientific concerns, are dismissed as "perceptions." As in, "There is a perception that the agricultural industry is only interested in profit, even at the risk of imposing damage or risk to human health. The task is to better manage this process so agricultural biotechnology is not at risk due to these perceptions." (Columnist Barb Sanderson in *Western Producer*, 9/7/98)

Precautionary Principle: Look before you leap - and if you are not sure of what you see, don't!

Protection: As in 1) Crop protection: technology protection systems (TPS - Terminator); 2) Corp protection: protecting corporate profits.

Questioning: Unhealthy for career advancement in the biotech sector.

Reductionist: "The process of reducing objects or organisms to their smallest parts rather than looking at them as a whole...In biology reductionism fosters the belief that the behavior of an organism or a tissue can best be explained by studying its cells, molecules and atoms and describing their constitution and function." (Hubbard & Wald, *Exploding the Gene Myth*)

Rights: Appear in a variety of forms, such as "ownership rights": "Pioneer Hi-Bred will take aggressive steps to protect its ownership rights for its germplasm." (company press release)

Risk Analysis: Academic gambling exercise.

Risks: Possible, but not probable, negative events; inherently a matter of odds, not to be taken seriously. "Some of the potential risks [of genetic engineering] are almost impossible to predict." (Positive Statement of 2/4/98 of UK Institute of Food Science & Technology on Genetic Modification and Food). The issue is redefined as "risk management" and we are assured that the industry and its promoters will manage all possible risks. There are no contingency plans.

Safety: A reductionist concept applied to food as a compilation of known component parts or ingredients. Food safety refers specifically to the absence of toxins known for their "catastrophic" effects (or, in quantities large enough to have immediate — and immediately noticeable — detrimental effect on personal health). Closely related to the concept of risk analysis.

Seeds: 1) Envelopes of genetics; 2) Not what they used to be.

Sound Science: 1) Listen to the music; 2) The art of harmonization, as in trade agreements.

Substantial Equivalence: For regulatory purposes, a product to be sold as a food does not need to be regulated if it is substantially equivalent to a familiar food (see "familiarity"). For example, a product that is red, thin skinned, perhaps juicy, and may weigh anywhere from a few ounces to a pound is substantially equivalent to a tomato.

Technology: Euphemism for genetic engineering.

Unknown: Traditional/indigenous knowledge of plants, animals, and ecology "discovered" and patented by drug companies.

Variety: A fuzzy taxonomic classification except when patented.

Whole: A concept foreign to genetic engineering.

X: Warning symbol applied to genetically engineered crops and foods.

Appendix 2
The Science of Genetic Engineering

We find it mixed in our food on the shelves in the supermarket — genetically engineered soybeans and maize. We find it growing in a plot down the lane — test field release sites with genetically engineered rape seed, sugar beet, wheat, potato, strawberries and more. There has been no warning and no consultation.

It is variously known as genetic engineering, genetic modification or genetic manipulation. All three terms mean the same thing, the reshuffling of genes usually from one species to another; existing examples include: from fish to tomato or from human to pig. Genetic engineering (GE) comes under the broad heading of biotechnology.

But how does it work? If you want to understand genetic engineering it is best to start with some basic biology.

What is a cell?

A cell is the smallest living unit, the basic structural and functional unit of all living matter, whether that is a plant, an animal or a fungus.

Some organisms such as amoebae, bacteria, some algae and fungi are single-celled — the entire organism is contained in just one cell. Humans are quite different and are made up of approximately 3 million million cells — (3,000,000,000,000 cells).

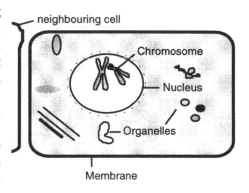

Cells can take many shapes depending on their function, but commonly they will look like a brick with rounded corners or an angular blob — a building block.

Cells are stacked together to make up tissues, organs or structures (brain, liver, bones, skin, leaves, fruit, etc.).

In an organism, cells depend on each other to perform various functions and tasks; some cells will produce enzymes, others will store sugars or fat; different cells again will build the skeleton or be in charge of communication like nerve cells; others are there for defence, such as white blood cells or stinging cells in jelly fish and plants.

In order to be a fully functional part of the whole, most cells have got the same information and resources and the same basic equipment.

A cell belonging to higher organisms (e.g. plant or animal) is composed of —

  • a cell MEMBRANE enclosing the whole cell. (Plant cells have an additional cell wall for structural reinforcement.)

  • many ORGANELLES, which are functional components equivalent to the organs in the body of an animal e.g. for digestion, storage, excretion.

  • a NUCLEUS, the command centre of the cell. It contains all the vital information needed by the cell or the whole organism to function, grow and reproduce. This information is stored in the form of a genetic code on the chromosomes that are situated inside the nucleus.

Proteins

Proteins are the basic building materials of a cell, made by the cell itself. Looking at them in close-up they consist of a chain of *amino-acids*, small specific building blocks that easily link up. Though the basic structure of proteins is linear, they are usually folded and folded again into complex structures. Different proteins have different functions. They can be transport molecules (e.g. oxygen-binding *haemoglobin* of the red blood cells); they can be antibodies, messengers, enzymes (e.g. digestion enzymes), or hormones (e.g. growth hormones or insulin). Another group is the structural proteins, which form boundaries and provide movement, elasticity and the ability for contraction. Muscle fibres, for example, are mainly made of proteins.

Proteins are thus crucial in the formation of cells and in giving cells the capacity to function properly.

Chromosomes

Chromosomes means "colored bodies" (they can be seen under the light microscope, using a particular stain). They look like bundled up knots and loops of a long thin thread. Chromosomes are the storage place for all genetic — that is hereditary — information. This information is written along the thin thread, called **DNA**. "DNA" is an abbreviation for *deoxyribo nucleic acid*, a specific acidic material that can be found in the *nucleus*. The genetic information is written in the form of a code, almost like a music tape. To ensure the thread and the information are stable and safe, a twisted double stranded thread is used, the famous *double helix*. When a cell multiplies it will also copy all the DNA and pass it on to the daughter cell. The totality of the genetic information of an organism is called *genome*.

Cells of humans, for example, possess two sets of 23 different chromosomes, one set from the mother and the other from the father. The

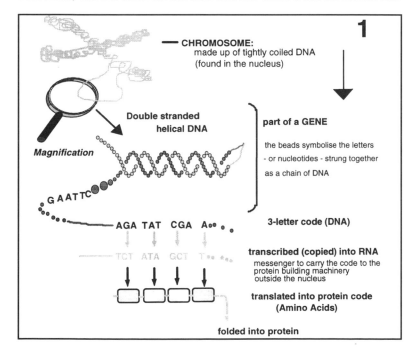

DNA of each human cell corresponds to 2 metres of DNA if it is stretched out and it is thus crucial to organize the DNA in chromosomes, so as to avoid knots, tangles and breakages. The length of DNA contained in the human body is approximately 60,000,000,000 kilometres. This is equivalent to the distance to the moon and back 8000 times!

The information contained on the chromosomes in the DNA is written and coded in such a way that it can be understood by almost all living species on earth. It is thus termed the *universal code of life.*

In this coding system, cells need only four symbols (called *nucleotides*) to spell out all the instructions of how to make any *protein.* Nucleotides are the units DNA is composed of and their individual names are commonly abbreviated to the letters A, C, G and T. These letters are arranged in 3-letter words, which in turn code for a particular *amino acid* — as shown in the flow diagram 1. The information for how any cell is structured or how it functions is all encoded in single and distinct **genes**. A **Gene** is a certain segment (length) of DNA with specific instructions for the production of a specific protein. The coding sequence of a gene is on average about 1000 letters long. Genes code for example for insulin, digestive enzymes, blood clotting proteins, or pigments.

The Regulation of Gene Expression:

How does a cell know when to produce which protein and how much of it?

In front of each gene there is a stretch of DNA that contains the regulatory elements for that specific gene, most of which is known as *the promoter*. It functions like a "control tower," con-

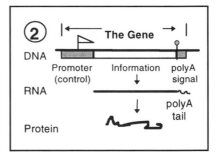

stantly holding a "flag" up for the gene it controls. Take insulin production (which we produce to enable the burning of the blood sugar) for example. When a message arrives in the form of a molecule that says "more insulin," the insulin control tower will signal the location of the insulin gene and say "over here." The message molecule will "dock in" and thus activate a "switch" to start the whole process of gene expression.

How does the information contained in the DNA get turned into a protein at the right time? As shown in picture 2, each gene consists of 3 main components: a "control tower" *(promoter)*, an information block and a polyA signal element.

If there is not enough of a specific protein present in the cell, a message will be sent into the nucleus to find the relevant gene. If the control tower recognizes the message as valid it will open the "gate" to the information block. Immediately the information is copied —or *transcribed*—into a threadlike molecule, called **RNA**. RNA is very similar to DNA, except it is single stranded. After the copy is made, a string of up to 200 "A"-type nucleotides —a polyA tail—is added to its end (picture 2). This process is called *poly-adenylation* and is initiated by a polyA signal located towards the end of the gene. A *polyA* tail is thought to stabilize the RNA message against degradation for a limited time. Now the RNA copies of the gene exit the nucleus and get distributed within the cell to little work units that translate the information into proteins.

No cell will ever make use of all the information coded in its DNA. Cells divide the work up amongst one another - they specialize. Brain cells will not produce insulin, liver cells will not produce saliva, nor will skin cells start producing bone. If they did, our bodies would be chaos!

The same is true for plants: root cells will not produce the green chlorophyll, nor will the leaves produce pollen or nectar. Furthermore, expression is age dependent: young shoots will not express any genes to do with fruit ripening, while old people will not usually start developing another set of teeth (exceptions have been known).

All in all, gene regulation is very specific to the environment in which the cell finds itself and is also linked to the developmental stages of an organism. So if I want the leaves of poppy plants to produce the red color of the flower petals I will not be able to do so by traditional breeding methods, despite the fact that leaf cells will have all the genetic information necessary. There is a block which prevents the leaves from going red. This block can be caused by two things:

• The "red" gene has been permanently shut down and bundled up thoroughly in all leaf cells. Thus the information cannot be accessed any more.

• The leaf cells do not need the color red and thus do not request RNA copies of this information. Therefore no message molecule is docking at the "red" control tower to activate the gene.

Of course—you might have guessed—there is a trick to fool the plant and make it turn red against its own will. We can bring the red gene in like a Trojan horse, hidden behind the control tower of a different gene. But for this we need to cut the genes up and glue them together in a different form. This is where breeding ends and genetic engineering begins.

Breeding

Breeding is the natural process of sexual reproduction within the same species. The hereditary information of both parents is combined and passed on to the offspring. In this process the same sections of DNA can be exchanged between the same chromosomes, but genes will always remain at their very own and precise position and order on the chromosomes. A gene will thus always be surrounded by the same DNA unless mutations or accidents occur. Species that are closely related might be able to interbreed, like a donkey and a horse, but their offspring will usually be infertile (e.g. mule). This is a natural safety device, preventing the mixing of genes that might not be compatible and to secure the survival of the species.

Genetic Engineering

Genetic engineering (GE) is used to take genes and segments of DNA from one species, e.g. fish, and put them into another species, e.g. tomato. To do so, GE provides a set of techniques to cut DNA either randomly or at a number of specific sites. Once they are isolated, one can study the different segments of DNA, multiply them up and *splice* them (stick them) next to any other DNA of another cell or organism. GE makes it possible to break through the species barrier and to shuffle information between completely unrelated species; for example, to *splice* the anti-freeze gene from flounder into tomatoes or strawberries, an insect-killing toxin gene from bacteria into maize, cotton or rape seed, or genes from humans into pig.

Yet there is a problem—a fish gene will not work in a tomato unless I give it a promoter with a "flag" the tomato cells will recognize. Such a

---

How to get the gene into the other cell

There are different ways to get a gene from A to B or to *transform* a plant with a "new" gene. A vector is something that can carry the gene into the host, or rather into the nucleus of a host cell (a). Vectors are commonly bacterial *plasmids* (see below) or viruses. Another method is the "shotgun technique," also known as "bio-ballistics," which blindly shoots masses of tiny gold particles coated with the gene into a plate of plant cells, hoping to land a hit somewhere in the cell's DNA (b).

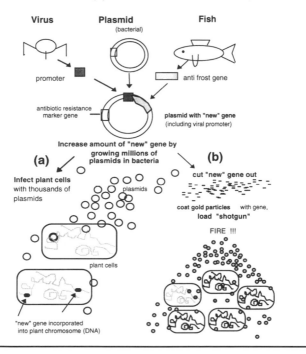

---

control sequence has either to be a tomato sequence or something else. Most companies and scientists do a shortcut here and don't even bother to look for an appropriate tomato promoter, as it would take years to understand how the cell's internal communication and regulation works. In order to avoid long testing and adjusting, most genetic engineering of plants is done with *viral promoters*. Viruses—as you will be aware—are very active. Nothing, or almost nothing, will stop them once they have found a new victim or rather *host*. They integrate their genetic information into the DNA of a host cell (such as one of your own), multiply, infect the next cells and multiply. This is possible because viruses have evolved very powerful promoters that command the host cell to constantly read the viral genes and produce viral proteins. Simply by taking a control element *(promoter)* from a plant virus and sticking it in front of the information block of the fish gene, you can get this combined virus/fish gene (known as a "construct") to work wherever and whenever you want in a plant.

This might sound great, the drawback though is that it can't be stopped either, it can't be switched off. The plant no longer has a say in the expression of the new gene, even when the constant involuntary production of the "new" product is weakening the plant's defences or growth.

And furthermore, the theory doesn't hold up with reality. Often, for no apparent reason, the new gene only works for a limited amount of time and then falls silent. But there is no way to know in advance if this will happen. Though often hailed as a precise method, the final stage of placing the new gene into a receiving higher organism is rather crude, seriously lacking both precision and predictability. The "new" gene can end up anywhere, next to any gene or even within another gene, disturbing its function or regulation. If the "new" gene gets into the "quiet" non-expressed areas of the cell's DNA, it is likely to interfere with the regulation of gene expression of the whole region. It could potentially cause genes in the "quiet" DNA to become active.

Often genetic engineering will not only use the information of one gene and put it behind the promoter of another gene, but will also take bits and pieces from other genes and other species. Although this is aimed to benefit the expression and function of the "new" gene it also causes more interference and enhances the risks of unpredictable effects.

What is a plasmid?

PLASMIDs can be found in many bacteria and are small rings of DNA with a limited number of genes. Plasmids are not essential for the survival of bacteria but can make life a lot easier for them. Whilst all bacteria—no matter which species—will have their bacterial chromosome with all the crucial hereditary information of how to survive and multiply, they invented a tool to exchange information rapidly. If one likens the chromosome to a bookshelf with manuals and handbooks, and a single gene to a recipe or a specific building instruction, a plasmid could be seen as a pamphlet. Plasmids self-replicate and are thus easily reproduced and passed around. Plasmids often contain genes for antibiotic resistance. This type of information, which can easily be passed on, can be crucial to bacterial strains that are under attack by drugs and is indeed a major reason for the quick spread of antibiotic resistance.

Working with plasmids

Plasmids are relatively small, replicate very quickly and are thus easy to study and to manipulate. It is easy to determine the sequence of its DNA, that is, finding out the sequence of the letters (A, C, G and T) and numbering them. Certain letter combinations—such as CAATTG—are easy to cut with the help of specific enzymes (see proteins). These cutting enzymes, called "restriction enzymes," are part of the Genetic Engineering "tool-kit" of biochemists. So if I want to *splice* a gene from fish into a plasmid, I have to take the following steps: I place a rather large number of a known plasmid in a little test tube and add a specific cutting enzyme; after an hour or so I stop the digest, purify the cut plasmid DNA and mix it with copies of the fish gene; after some time the fish gene places itself into the cut ring of the plasmid. I quickly add some "glue" from my "tool-kit"—an enzyme called *ligase*—and place the mended plasmids back into bacteria, leaving them to grow and multiply. Of course my plasmid would also have its own genes, or genes I would have placed there beforehand in order to ease my experiments. The genes crucial for a smooth process are the **MARKER GENES**. These marker genes are commonly genes for antibiotic resistance. If a plasmid is marked with a gene for antibiotic resistance I can now add the specific antibiotic to the food supply of the bacteria. All those that do not have the plasmid will die, and all those that do have the plasmid will multiply.

What's wrong with Genetic Engineering?

Genetic Engineering is a test tube science and is prematurely applied in food production. A gene studied in a test tube can only tell what this gene does and how it behaves in that particular test tube. It cannot tell us what its role and behavior are in the organism it came from or what it might do if we place it into a completely different species. Genes for the color red placed into petunia flowers not only changed the color of the petals but also decreased fertility and altered the growth of the roots and leaves. Salmon genetically engineered with a growth hormone gene not only grew too big too fast but also turned green. These are unpredictable side effects, scientifically termed *pleiotropic effects.*

We also know very little about what a gene (or for that matter any of its DNA sequence) might trigger or interrupt depending on where it got inserted into the new host (plant or animal). These are open questions around *positional effects*. And what about **gene silencing** and **gene instability?**

How do we know that a genetically engineered food plant will not produce new toxins and allergenic substances or increase the level of dormant toxins and allergens? How about the nutritional value? And what are the effects on the environment and on wild life? All these questions are important questions yet they remain unanswered. Until we have an answer to all of these, genetic engineering should be kept to the test tubes. Biotechnology married to corporations tends to ignore the precautionary principle but it also ignores some basic scientific principles.

Researched and written by Dr. Ricarda Steinbrecher, with graphics by Faye Kenner, for the Women's Environmental Network (WEN). The WEN Trust is one of Britain's leading environmental charities. Its aim is to educate, inform, and empower women who care about the environment.

Recommended Reading

Anderson, Robert , Edwin Levy, and Barrie Morrison. *Rice Science and Development Politics - IRRI's Strategies and Asian Diversity 1950-1980.* Oxford: Clarendon Press, 1991.

Bud, Robert. *The Uses of Life: A History of Biotechnology.* Cambridge: Cambridge University Press, 1993.

Busch, Larry, Wm. Lacey, Jeffery Burkhardt, and Laura Lacy. *Plants, Power, and Profit: Social, Economic and Ethical Consequences of the New Biotechnologies.* Oxford: Blackwell, 1991.

Fowler, Cary, *Unnatural Selection: Technology, politics and Plant Evolution.* Amsterdam: Gordon and Breach, 1994.

Fowler, Cary, and Pat Mooney. *Shattering: Food, Politics, and the Loss of Genetic Diversity.* Tucson: University of Arizona Press, 1990.

Grace, Eric. *Biotechnology Unzipped.* Toronto: Trifolium Books, 1997.

Harding, Sandra. *The Science Question in Feminism.* Ithaca: Cornell University Press, 1986.

Ho, Mae-Wan. *Genetic Engineering: Dream or Nightmare.* New Delhi: Research Foundation for Science, Technology & Ecology, 1997 / Penang: Third World Network, 1998.

Holdrege, Craig. *Genetics and the Manipulation of Life: The Forgotten Factor of Context.* Hudson, NY: Lindisfarne Press, 1996.

Hubbard, Ruth, and Elijah Wald. *Exploding the Gene Myth.* Boston: Beacon Press, 1993.

Juma, Calestous. *The Gene Hunters.* Princeton: Princeton University Press, 1989.

Keller, Evelyn Fox. *A Feeling for the Organism: The Life and Work of Barbara McClintock.* New York: Freeman, 1983.

Keller, Evelyn Fox. *Reflections on Gender and Science.* New Haven: Yale University Press, 1985.

Keller, Evelyn Fox. *Secrets of Life, Secrets of Death: Essays on Language, Gender and Science.* New York & London: Routledge, 1992.

Keller, Evelyn Fox. *Refiguring Life: Metaphors of Twentieth-Century Biology.* New York: Columbia University Press, 1995.

Kloppenburg, J.R. *First the Seed: The Political Economy of Plant Biotechnology.* Cambridge: Cambridge University Press, 1988.

Kneen, Brewster. *The Rape of Canola.* Toronto: NC Press, 1993.

Krimsky, Sheldon. *Biotechnics and Society: The Rise of Industrial Genetics.* New York: Praeger, 1991.

Lappé, Marc. *Broken Code; The Exploitation of DNA.* San Francisco: Sierra Club, 1984.

Lappé, Marc. *Evolutionary Medicine: Rethinking the Origins of Disease.* San Francisco: Sierra Club, 1994.

Latour, Bruno. *We Have Never Been Modern.* Cambridge: Harvard University Press, 1993.

Levins, Richard, and Richard Lewontin. *The Dialectical Biologist.* Cambridge: Harvard University Press, 1985.

Lewontin, R.C. *Biology as Ideology: The Doctrine of DNA* (Massey Lectures). Toronto: Anansi, 1991.

Lewontin, R.C., Steven Rose, and Leon Kamin. *Not In Our Genes: Biology, Ideology, and Human Nature.* New York: Pantheon, 1984.

Mander, Jerry. *In the Absence of the Sacred: The Failure of Technology and the Survival of the Indian Nations.* San Francisco: Sierra Club, 1991.

Maranto, Gina. *Quest for Perfection: The Drive to Breed Better Human Beings.* New York: Scribner, 1996.

Mies, Maria, and Vandana Shiva. *Ecofeminism.* London: Zed Books, 1993.

Nabhan, Gary. Enduring Seeds: *Native American Agriculture and Wild Plant Conservation.* San Francisco: North Point Press, 1989.

Nottingham, Stephen. *Eat Your Genes.* London: Zed Books, 1998.

Rose, Steven. *Lifelines: Biology Beyond Determinism.* Oxford: Oxford University Press, 1998.

Shiva, Vandana. *The Violence of the Green Revolution.* London: Zed Books, 1991.

Shiva, Vandana. *Monocultures of the Mind.* London: Zed Books, 1993.

Shiva, Vandana, and Ingunn Moser. *Biopolitics: A Feminist and Ecological Reader on Biotechnology.* London: Zed Books, 1995.

Steingraber, Sandra. *Living Downstream: An Ecologist Looks at Cancer and the Environment*. Reading, MA: Addison Wesley, 1997.

Turney, Jon. *Frankenstein's Footsteps: Science, Genetics and Popular Culture*. New Haven: Yale University Press, 1998.

van Dommelen, Ad, ed. *Coping with Deliberate Release: The Limits of Risk Assessment*. Tilburg, Netherlands: International Centre for Human and Public Affairs, 1996.

Winner, Langdon. *The Whale and the Reactor: A Search for Limits in an Era of High Technology*. Chicago: University of Chicago Press, 1986.

Yoxen, Edward. *The Gene Business: Who Should Control Biology?* New York: Harper and Row, 1983.

Yoxen, Edward. *Unnatural Selection? Coming to Terms with the New Genetics*. London: Heinemann, 1986.

Periodicals:

The Ram's Horn (monthly, $20/year)

S-6, C-27, RR.1 Sorrento, BC, V0E 2W0 ramshorn@jetstream.net

Biotechnology & Development Monitor (quarterly, free)

University of Amsterdam, Department of Political Science Oudezijds Achterburgwal 237 1012 DL Amsterdam, The Netherlands

Seedling (quarterly) (Free to NGOs and to groups/individuals in the south. US$25 to institutions and others in industrialized countries) GRAIN (Genetic Resources Action International) Girona 25, pral. E-8010 Barcelona, Spain www.grain.org

RAFI Communiqué (free/donation)

Rural Advancement Fund International 110 Osborne St., Suite 202 Winnipeg, MB R3L 1Y5, Canada rafi@rafi.org www.rafi.ca

Notes

Introduction

1 Months after writing this I found this sentence in an editorial from a 1972 issue of the *Journal of the American Medical Association* in 1972: "Thus the earliest procedure in genetic engineering might be considered to be artificial insemination," quoted in Jon Turney, *Frankenstein's Footsteps: Science, Genetics and Popular Culture* (New Haven: Yale University Press, 1998)

2 Keith Betteridge, "Bovine Embryo Research" in *Agri-food research in Ontario*, Guelph, September 1995

3 *Wisconsin State Farmer*, 9/8/96; *Nature Biotechnology*, May 1998

4 Brewster Kneen, *The Rape of Canola* (Toronto: NC Press, 1992)

Chapter 1

5 For information about this subject, contact Pesticides Action Network North America Regional Centre, 49 Powell Street #500, San Francisco, CA 94102, panna@panna.org

6 For characteristics of a healthy, living soil, see www.soilfoodweb.com

7 *New Scientist*, 25/7/98; *Science*, vol. 281, p. 428

8 Lee Hoinacki, unpublished manuscript, 15/7/95

9 Harry Collins, vice president D&PL, "New Technology and Modernizing World Agriculture," report distributed to FAO delegates in Rome, June 1998

10 Rural Advancement Fund International (RAFI), e-mail communication, 11/3/98

Chapter 2

11 US Department of Agriculture press release, 18/11/97

12 Laura Rance, *Manitoba Co-operator*, 9/7/98

13 George G. Khachatourians, Department of Applied Microbiology, University of Saskatchewan, writing in the *AgBiotech Bulletin*, February 1998. In May 1998 Khachatourians was appointed to the Standards/Certification Task Force of BioteCanada, the industry lobby organization.

14 Maria Mies and Vandana Shiva, *Ecofeminism* (London: Zed Books, 1993), p. 175

15 David Cayley, *Ivan Illich in Conversation* (Toronto: Anansi, 1992), p. 270

16 "Our Precious Planet" in *Time*, November 1997

17 Gregg Easterbrook, "Forgotten Benefactor of Humanity" in *Atlantic Monthly*, January 1997

18 Norman Borlaug , "Feeding a World of 10 Billion People: The Miracle

Ahead," a lecture at De Montfort University, Leicester, UK, May 6, 1997
19 Alan Irwin, *Citizen Science*, (London/New York: Routledge, 1994), pp. 4, 14
20 *Milling & Baking News*, 8/4/97
21 *Ottawa Citizen*, 3/3/98
22 *New Scientist*, 11/4/98
23 Mae-Wan Ho, *Genetic Engineering: Dreams or Nightmares?* (New Delhi: Research Foundation for Science, Technology and Ecology/Third World Network, 1998), p. 115
24 "Monsanto v. Malthus" in *Forbes*, 10/3/97
25 *World Population Projections to 2150* (New York: UN Population Division, 1998)
26 *New Scientist*, 20/6/98
27 *New Scientist*, 11/7/98

Chapter 3
28 Ivan Illich, "The Institutional Construction of a New Fetish: Human Life," presented at a planning event of the Evangelical Lutheran Church in America, Chicago, 29/3/89, and reprinted in Brewster Kneen, Ivan Illich,and the EECCS Bioethics Working Group, *Drawing the Line: The Ethics of Biotechnology*, Occasional Paper #5 (Brussels: Ecumenical Association for Church and Society, 1997)
29 Quoted by Richard Doyle in *On Beyond Living: Rhetorical Transformations of the Life Sciences* (Stanford: Stanford University Press, 1997), p. 10
30 ibid.
31 Steven Rose, *Lifelines* (Oxford: Oxford University Press, 1997), p. ix
32 Jonathan Ramsey, information manager, Monsanto Europe, in *Chemistry and Industry*, No. 14, 20/7/98, pp. 549-584
33 *Business Ethics*, Jan/Feb 1996
34 Godfrey B. Tangwa, University of Yaounde, Cameroon, *Monitor* #32 (Amsterdam), September 1997
35 *TOMORROW - Global Environment Business* (Sweden), March-April 1997
36 Teruhiko Waskayama and Ryuzo Yanagimachi, *Nature Biotechnology*, July 1998, p. 639
37 Lovell-Badge, *Nature Biotechnology*, July 1998, p. 618
38 *Farm & Country* (Toronto), 7/9/98
39 Jon Turney, *Frankenstein's Footsteps*, p. 48
40 See Wm Cronon, *Changes in the Land* (New York: Hill & Wang, 1983)
41 Evelyn Fox Keller, *A Feeling for the Organism: The Life and Work of Barbara McClintock* (New York: Freeman, 1983), p. 180
42 ibid., p. 158
43 ibid., p. 103
44 ibid.
45 *Nature Biotechnology*, June 1998
46 Frédérique Apffel-Marglin, "Counter-Development in the Andes" in *The Ecologist*, November/December 1997 (Abridged from original in *InterCulture* #126, winter 1995)

Chapter 4

47 "Sociopolitical Effects of New Biotechnologies in Developing Countries," a paper from the International Food Policy Research Institute, Washington, 1995

48 K. Dohmen, *Gentechnologie — Die andere Schspfung?* (Stuttgart: Metzler, 1988)

49 AAFC, "Sustainable Development Strategy for the Auditor General," Ottawa, 1997

50 "Let the Harvest Begin," posted on http://www.monsanto.com, 14/10/98

51 "How the Terminator Terminates," an occasional paper of The Edmonds Institute, Edmonds, Washington, 1998

52 Sandra Harding, *The Science Question in Feminism* (Ithaca: Cornell University Press, 1986), p. 16

53 Robert Bud, *The Uses of Life: A History of Biotechnology* (Cambridge: Cambridge University Press, 1993), pp. 41-42

54 ibid., p. 103

55 ibid., pp. 121-124

56 ibid., p. 126

57 Attributed to Ronald Cape of Cetus Corp. by John Elkington in *The Gene Factory* (New York: Carroll & Graf, 1985)

58 Bud, *Uses of Life*, p. 164

59 Government of Canada news release 6/8/98

60 Keller, *A Feeling for the Organism*, p. 160

61 Rose, *Lifelines*, p. 273

62 Turney, *Frankenstein's Footsteps*, pp. 119-120

63 Rose, *Lifelines*, pp. 273-4

64 Philip Regal, "Metaphysics in Genetic Engineering: Cryptic Philosophy and Ideology in the 'Science' of Risk Assessment" in Ad van Dommelen, ed., *Coping with Deliberate Release: The Limits of Risk Assessment* (Tilburg, Netherlands: International Centre for Human and Public Affairs, 1996), pp. 18, 19

65 Robert S. Anderson, Edwin Levy, and Barrie Morrison, *Rice Science and Development Politics: IRRI's Strategies and Asian Diversity 1950-1980* (Oxford: Clarendon Press, 1991), p. 22

66 ibid., p. 30

67 Turney, *Frankenstein's Footsteps*, pp. 189-190

68 ibid., p. 191

69 Fred Buttel, "Ideology and Agricultural Technology in the Late Twentieth Century" in *Agriculture and Human Values*, spring 1993, p. 9

70 ibid.

71 ibid., p. 10

Chapter 5

72 Ho, *Genetic Engineering*, p. 179

73 See Jack Kloppenberg, *First the Seed: The Political Economy of Plant Biotechnology* (Cambridge: Cambridge University Press, 1988)

74 Posted at http://www.cargill.com

75 *Western Producer* (Saskatoon) 18/6/98
76 RAFI news releases, 11/3/98, 13/3/98

Chapter 6
77 Reuters, 22/6/98
78 A grant from the Social Sciences and Humanities Research Council of Canada for work as a "private scholar," i.e., not a university employee, made it possible.
79 Science Council of Canada, "Seeds of Renewal," Report 38, Ottawa, 1985
80 Agriculture Canada, "Commodity Strategy Framework," Ottawa, September 1986, p. 88
81 J. Grant Smith, chairman, Ontario Milk Marketing Board, presentation at the 1988 Annual Meeting of the OMMB. It was Smith's herd that was secretly under contract to Eli Lilly for research purposes.
82 Dairy Famers of Canada, "Somatotropin (Bovine Growth Hormone) and the Dairy Industry," Ottawa, May 1987
83 Robert Kalter et al, *Biotechnology and the Dairy Industry,* Agricultural Economics Research 85-20, Cornell University, Ithaca, 1985
84 *Proceedings,* USDA National Invitational Workshop on Bovine Somatatrophin, St. Louis, MO, 1987, pp. 46-47
85 ibid., pp. 48-49
86 Tess Hooks, "The Political Economy of Industry University Relationships and Economic Change: A Case Study" (master's thesis, Department of Sociology, Cornell University, 1988)
87 Tess Hooks is at the Department of Sociology, University of Western Ontario, London, Ontario; Frederick H. Buttel is at the Department of Rural Sociology, University of Wisconsin, Madison, Wisconsin; David S. Kronfeld is at the Department of Animal and Poultry Sciences and the Department of Large Animal Clinical Sciences, Virginia Polytechnic Institute and State University, Blacksburg Virginia.
88 *New Scientist,* 11/2/88
89 The author has copies of this letter and other documents cited here on file
90 Notes, May 17, 1989
91 Notes on a meeting of the NMPF BST Advisory Committee, July 7, 1989
92 Judith Juskevich and C. Greg Guyer, "Bovine Growth Hormone: Human Food Safety Evaluation," *Science,* August 24, 1990
93 Wade Roush, "Who Decides about Biotech," *Technology Review,* July 1991
94 ibid.
95 ibid.
96 BGH News of the Week, newsletter from Institute for Agriculture and Trade Policy (IATP), 10/11/94
97 Roush, ibid.
98 *Science,* August 1998, p. 933
99 Paul Brodeur and Bill Ravanesi, "Living Downstream: The New England Journal of Medicine and Conflict of Interest," *The Networker,* June 1998 (electronic publication) The article fully documents the behavior of W.R. Grace and the *New England Journal of Medicine.*

100 *New Scientist*, 25/7/98
101 Research Report, Alberta Agriculture, October 1986
102 *Forestry & Agriculture Bulletin*, summer 1988
103 Brian McBride, interview, January 12, 1988
104 Jeanne Burton and Brian McBride, "Recombinant Bovine Somatotropin (rbST): Is There a Limit for Biotechnology in Applied Animal Agriculture?" *Journal of Agricultural Ethics*, Vol. 2, #2, 1989, pp. 129, 130
105 Brian McBride, interview, January 1988
106 John Unhoefer, *Dairy Foods*, April 1987
107 *Western Producer*, 5/9/96
108 *Globe & Mail*, 8/5/98
109 *Nature Biotechnology*, March 1998
110 Posted at http://www.monsanto.com
111 *Globe & Mail*, 26/1/98
112 National Research Council of Canada, *Biotechnology Bulletin*, Ottawa, June 1990

Chapter 7
113 *Comstock's*, June 1991
114 Calgene stock prospectus, 1989 and 1992
115 *San Francisco Examiner*, 10/1/93
116 *New York Times*, 12/1/93
117 *The Packer*, 7/8/93
118 *The Packer*, 18/4/94
119 *Wall Street Journal*, 29/6/95
120 Monsanto annual report, 1995
121 Calgene annual report, 1994

Chapter 8
122 Robert Plaisted, phone interview, 19/10/98
123 *New Scientist*, 28/8/93
124 *New Scientist*, 28/8/93
125 EPA press release, 11/8/95
126 Agriculture Canada Decision Document 96-06, "Determination of Environmental Safety of NatureMark Potatoes," Ottawa, 1996
127 Rose, *Lifelines*, p. 166
128 Mycogen-UCS e-mail exchange, posted by the Union of Concerned Scientists, 27/6/95
129 *Ontario Farmer*, 5/3/96
130 *Ontario Farmer*, 17/12/96
131 *Ontario Farmer*, 17/9/96
132 *Ontario Farmer*, 27/5/97
133 Pioneer Hi-Bred Ltd., personal communication, 15/9/98
134 Ameriscan, 28/9/98. In the report the National Corn Growers Association reports that its "membership includes over 30,000 growers in 48 states and 44 affiliated state corn grower organizations. NCGA's mission is to create and increase opportunities for corn growers in a changing world and

enhance corn utilization and profitability."
135 *Ontario Farmer*, 29/9/98
136 *New Scientist*, 21/2/98
137 Received via e-mail distribution
138 Cornell University News Service, 27/4/98
139 *New Scientist*, 22/8/98
140 *Globe & Mail*, 6/2/96
141 *Globe & Mail*, 9/2/96

Chapter 9
142 Richard Mahoney, then president of Monsanto, speaking at the Executive Club of Chicago, 12/2/93
143 Science Council of Canada, "Regulating the Regulators — Science, Values and Decisions," Ottawa, 1982, p. 70
144 *Meat & Poultry*, 10/98
145 Agriculture Canada, "Food Irradiation: An Alternative Food Processing Technology," Ottawa, 1989
146 *This Is Codex Alimentarius*, Rome, 1993 (2nd edition 1995)
147 Comments to the Codex Committee on Food Labelling, 26th session, May 1998
148 Bruno Latour, "Socrates' and Callicles' Settlement, or The Invention of the Impossible Body Politic," *Configurations* (Johns Hopkins University Press & the Society for Literature and Science), 5:189-240, 1997
149 National Biotechnology Advisory Committee, "Leading in the Next Millennium," Sixth Report, Ottawa, 1998
150 See Brewster Kneen, *Invisible Giant: Cargill and its Transnational Strategies* (London: Pluto Press, 1995)
151 Statement by Art Olson, president, CFIA, 2/5/97
152 Information bulletin, CFIA website, 1/4/97
153 *Manitoba Co-operator*, 26/2/98
154 Ekologiska Lantbrukarna (Swedish Association of Ecological Farmers) via genetnews, 13/3/98
155 Information bulletin, CFIA website, 1/4/97
156 Eric Kindberg, e-mail communication, 18/1/98
157 Science Council of Canada with the Institute for Research on Public Policy, *Biotechnology in Canada: Promises and Concerns*, Ottawa, 1980
158 Science Council of Canada, "Seeds of Renewal" Report 38, Ottawa, 1985
159 Canbiocon Symposium on the Regulation of Biotechnology, 19/9/89 (transcripts published May 1991)
160 Information letter, 1992
161 National Biotechnology Advisory Committee, Fifth Report, 1991
162 Health Protection Branch, Health and Welfare Canada, information letter (9 pages, bilingual), Ottawa, 5/8/92
163 *Codex Alimentarius*, Article One of the "Code of Ethics for International Trade in Foods," Rome, 1979 (revised 1985)
164 *Codex Alimentarius*, Articles 4.1, 4.2 of the General Principles of the "Code of Ethics"

165 Science Council of Canada, "Regulating the Regulators," 1982, p. 76

166 Office of Biotechnology, Canadian Food Inspection Agency, "The Safety-Based Approach to Regulation of Agricultural Products," Ottawa, May 1998

167 Mae-Wan Ho & Ricarda Steinbrecher, *Fatal Flaws in Food Safety Assessment: Critique of the Joint FAO/WHO Biotechnology and Food Safety Report (1996)* (Penang: Third World Network, 1998)

168 Nancy J. Myers, Carolyn Raffensperger, *The Networker*, June 1998 (electronic)

169 Quoted by Katherine Barrett and Carolyn Raffensperger in "Precautionary Science" in C. Raffensperger and J. Tickner (eds), *The Precautionary Principle: To Foresee and Forestall* (Island Press, in press)

170 Barrett and Raffensperger, ibid.

171 Laurie Curry, Food and Consumer Products Manufacturers of Canada, *Ontario Farmer Daily*, 30/9/98

172 *Biotechnology & Agriculture: A New Approach*, Agriculture Canada training video

173 Comments of the International Seed Trade Federation/International Association of Plant Breeders to the Codex Committee on Food Labelling, 26th session, May 1998

174 *Western Producer*, 20/11/97

175 *Western Producer*, 20/11/97

176 *Western Producer*, 22/1/98

177 ISB News (electronic newsletter), June 1997, citing: R.S. Hails, M. Rees, D. D. Kohn, and M. J. Crawley, "Burial and seed survival in Brassica napus subsp. oleifera and Sinapis arvensis including a comparison of transgenic and nontransgenic lines of the crop," *Proc. Royal Soc. Lond.* B 264:17, 1997; and, J. Bergelson, C. B. Purrington, C. J. Palm, and J.C. LopezGutierrez, "Costs of resistance: a test using transgenic Arabidopsis thaliana," *Proc. Royal Soc. Lond.* B 263:16591663, 1996

178 *Western Producer*, 3/9/98

179 Editorial in *New Scientist*, 4/7/98

180 *New Scientist*, 4/7/98

181 Bud, *Uses of Life*, p. 178

182 Ruth Hubbard and Elijah Wald, *Exploding the Gene Myth* (Boston: Beacon Press, 1993), p.117

183 ibid., p. 61

184 Rose, *Lifelines*, pp. 296-297

185 Hubbard & Wald, *Exploding the Gene Myth*, p. 69

186 *AgBiotech Bulletin*, May/June 1993

187 *AgBiotech Bulletin*, April 1997

188 Agriculture Canada, "Biotechnology in Agriculture - Science for Better Living" - no date

189 *Western Producer*, 9/1/97

190 *Western Producer*, 9/1/97

191 *Western Producer*, 10/9/98

Chapter 10
192 Rose, *Lifelines*, p. 171
193 Editorial, *Globe & Mail*, 8/8/98
194 From a speech given in Salzburg, Austria, 15/9/97, amended 2/10/97, and posted at http://www.monsanto.com
195 Keller, *A Feeling for the Organism*
196 Rose, *Lifelines*, p. 85
197 Craig Holdrege, *Genetics & the Manipulation of Life: The Forgotten Factor of Context* (Hudson, NY: Lindisfarne Press, 1996), p. 40
198 Turney, *Frankenstein's Footsteps*, p. 44, emphasis added
199 ibid., p. 48
200 Richard Strohman, letter to *Nature Biotechnology*, pp. 1224-1225, November 1997
201 Rose, *Lifelines*, p. 4
202 ibid., pp. 140-142
203 Nathan Rosenburg & L.E. Birdzell, "Science, Technology and the Western Miracle," *Scientific American*, November 1990
204 Evelyn Fox Keller, *Secrets of Life, Secrets of Death* (New York & London: Routledge, 1992), pp. 39-40
205 ibid., pp. 40-41
206 Mies, *Ecofeminism*, pp. 46-47. See also Shiva Vandana, *Staying Alive* (London: Zed Books, 1988)
207 Antonio Regalado, "The Troubled Hunt for the Ultimate Cell," *Technology Review*, July/August 1998
208 Keller, *Secrets*, p. 179
209 Richard Strohman, "The coming Kuhnian revolution in biology," *Nature Biotechnology*, March 1997, pp. 194-200
210 Antonia Regalado, "The Next Genome Project," *Technology Review*, May/June 1998
211 Hubbard & Wald, *Exploding the Gene Myth*, p. 52
212 *New Scientist*, 11/7/98
213 Maranto, *Quest for Perfection*, p. 24
214 *Globe & Mail*, 25/7/98
215 *Globe & Mail*, 6/2/99
216 *New Scientist*, 18/7/98
217 E.O. Wilson, *Consilience*, (New York: Knopf, 1998)

Chapter 11
218 David Noble, *The Religion of Technology* (New York: Knopf, 1997), p. 23
219 *The Economist*, 26/4/97
220 Turney, *Frankenstein's Footsteps*, p. 14
221 Ernst Benz, *Evolution and Christian Hope* (Garden City: Anchor Books, 1968), pp. 121-132
222 Keller, *Secrets*, p. 58
223 Benz, *Evolution*, p. 132
224 Bruno Latour, *We have Never Been Modern*, (Cambridge: Harvard University Press, 1993), pp. 33-34

225 Turney, *Frankenstein's Footsteps*, p. 27
226 C. Baldick, *In Frankenstein's Shadow: Myth, Monstrosity and Nineteenth Century Writing*, (Oxford: Oxford University Press, 1987), p. 4
227 Ho, *Genetic Engineering*, p. 71
228 Gary Nabhan, *Enduring Seeds*, (San Francisco: North Point Press, 1989), p. 74

Chapter 12
229 Albert Memmi, *The Colonizer and the Colonized* (Boston: Beacon Press, 1967)
230 *Guardian Weekly*, London, 4/10/98
231 Steve Emmott, advisor on biotechnology to the Green Party in the European Parliament, at the First Grassroots Gathering on Biodevastation, St. Louis, July 1998
232 *Samykta Karnataka*, Bangalore, 25/11/98
233 *Farmers' Weekly*, 5/2/99

Index

Numbers following the letter "n" refer to endnotes

About the Author

Brewster Kneen spent 15 years in Nova Scotia, Canada, building up a commercial sheep flock, two producer co-ops, and a monthly newsletter — *The Ram's Horn*, an analysis of the food system — which he and his wife continue to publish. He now lives in British Columbia's interior and uses his farm experience, along with his academic training in economics and theology, as he writes and lectures on issues in food and agriculture from barnyard, to boardroom, to biotechnology. He is known internationally as Canada's foremost expert on the global food system, and locally as a popular speaker whose wry wit tempers his uncompromising analysis of the nature and extent of transnational corporate control.

His books include *From Land to Mouth: Understanding the Food System* (1989) and the revised *Second Helping* (1993); *Trading Up — How Cargill, the World's Largest Grain Company, is Changing Canadian Agriculture* (1990); *The Rape of Canola* (1992); and *Invisible Giant: Cargill and its Transnational Strategies* (1995).

The Ram's Horn details the hidden wonders — and horrors — of the food system, and what people are doing about them. For example: the connections between the seed and agro-toxin companies, the ways in which agribusiness transnational corporations shape public policy, the genetic engineering of foods, community shared agriculture, and other visionary alternatives to the industrial system. It is available by subscription from:

The Ram's Horn, S-6, C-27, R.R. #1, Sorrento B.C., V0E 2W0, Canada.

℞